Contents

List of Contributors

A. G. Gordon
Partner, Forestart, Church Farm, Hadnall, Shrewsbury, Shropshire SY4 4AQ (formerly Principal Seeds Officer, Forestry Commission Research Division).

R. Faulkner
formerly Principal Geneticist, Forestry Commission, Northern Research Station, Roslin, Midlothian EH25 9SY.

R. Lines, OBE
formerly Silviculturist, Forestry Commission, Northern Research Station, Roslin, Midlothian EH25 9SY.

C. J. A. Samuel
Tree Improvement Branch, Forestry Commission, Northern Research Station, Roslin, Midlothian EH25 9SY.

J. R. Aldhous
formerly Head of Silviculture Division, Forestry Commission Headquarters, 231 Corstorphine Road, Edinburgh EH12 7AT.

A. M. Fletcher
Tree Improvement Branch, Forestry Commission, Northern Research Station, Roslin, Midlothian EH25 9SY.

Chapter 1

Introduction

by **A. G. Gordon**

Since the publication in 1972 of Forestry Commission Bulletin 43 *Nursery practice* there have been several significant developments, both statutory and experimental, which justify a more comprehensive publication devoted solely to seed. This Bulletin therefore comprises a series of chapters covering all phases of seed usage of commercial forestry species from source selection, through collection, processing, storage and legislation, to seed sowing; each chapter is written by individual specialists in their field.

In recent years substantial changes have taken place in the international forest tree seed market which have made the task of supplying British requirements from abroad rather less certain than previously. Although seed production from British forests has increased during the same period, most has come from Forestry Commission plantations. One object of this manual is to provide all the information necessary for organising and undertaking seed collections, and thus to stimulate greater seed collection efforts in the future throughout British forests.

The earlier the notice given that a seed crop is expected the more chance that any necessary changes to management plans can be made in order to harvest the crops. The earliest notice available is from flower bud formation and flowering. For this reason, and in the absence of any comprehensive coverage in any other readily available publications in Great Britain, Chapter 6 is devoted to a full description of the subject.

Forestry Commission Bulletin 66 *Choice of seed origins for the main forest species in Britain* gives a very complete account of all research to date on the subject. For the sake of completeness a full synopsis has been included in this manual as Chapter 3.

The principal species referred to in this manual are those covered by the Forest Reproductive Material Regulations 1977, together with *Betula* species, and some other important conifers, for example *Pinus contorta* and *Larix × eurolepis*, which are also fairly widely used for commercial forestry purposes in Great Britain. The restricted coverage of broadleaves is compensated for by the publication of Forestry Commission Bulletin 62 *Silviculture of broadleaved woodland* (1984).

This manual is a companion to Forestry Commission Bulletin 59 *Seed manual for ornamental trees and shrubs* which filled an important gap in British forestry literature. It is hoped that this manual will also provide the student and interested amateur, as well as the practical forester and nurseryman at whom it is mainly aimed, with a fuller background to the subject of seed collection and that it will act as a reference volume should they wish to pursue their interest more deeply.

Students should note that throughout most of the manual the terms fruit and seed are used in their broadest sense to describe respectively any aggregation of seeds such as cones, catkins, and any prime dispersal unit comprising food storage tissue, a latent growth axis and a protective covering. A comprehensive glossary of terms can be found at the end of this Bulletin (p.122).

REFERENCES

ALDHOUS, J.R. (Ed.) (1972). *Nursery practice.* Forestry Commission Bulletin 43. HMSO, London. (*Under revision.*).

ANON. (1987). *The forest reproductive material regulations 1977 – an explanatory booklet.* Forestry Commission, Edinburgh.

EVANS, J. (1984).*Silviculture of broadleaved woodland.* Forestry Commission Bulletin 62. HMSO, London.

GORDON, A.G. and ROWE, D.C.F. (1982). *Seed manual for ornamental trees and shrubs.* Forestry Commission Bulletin 59. HMSO, London.

LINES, R. (1987). *Choice of seed origins for the main forest species in Britain.* Forestry Commission Bulletin 66. HMSO, London.

Chapter 2

The Choices and Relative Values of Different Seed Sources

by **R. Faulkner**

Introduction

For most commercial species, home collections and importations are the only potential sources of seed available to nurserymen. With few exceptions, home collected seed of the best known origin should be used in preference to seed of the same origin imported either from countries where the species occurs naturally, or from third countries. During the growth of the parent tree crop in Britain, natural and silvicultural selection will have taken place, resulting in seed which is slightly improved genetically when compared with seed imported from abroad (Gill, 1983a). Plants raised from home collected seed will be better able to withstand British conditions in general – and in the locality of the seed stand in particular – than plants raised from imported seed. Not all home collected seed of common origin is necessarily of the same genetic quality, since genetic quality can be influenced to some extent by the intensity of human and environmental selection.

Several types of plantation for seed production exist; selected seed stands and seed orchards are the two principal kinds. The former are used for the large-scale production of seed of the best origins, whereas seed orchards are used for the production of seed of new and improved cultivars and hybrids.

All seed sources should normally be more than 4 ha in area and should be reasonably isolated by distance against contamination from undesirable pollen sources, for example, slow-growing stands with poor stem form, or other compatible species which readily form natural hybrids. Single trees and small groups of trees should never be used as a seed source for commercial forestry work because inbreeding risks are high and the consequences are often low seed set and seedlings which develop into crops of inferior quality.

The Forestry Commission is responsible for the selection and registration of suitable seed sources in Britain (see Chapter 5).

Vegetative propagation of stock of superior genetic quality

Vegetatively multiplied Sitka spruce planting stock of very high genetic quality first became available in limited quantities in Britain in 1983: supplies of *Larix × eurolepis* should become available in the early 1990s. Although outside the scope of this manual, mention of vegetative propagation is made for the sake of completeness. Vegetative propagation techniques provide the fastest method of introducing the most superior genetically improved stock at any given time into commercial plantations. Rooted cuttings offer great potential as they may be multiplied quite quickly from small quantities of seed, which may themselves have been produced artificially by controlled crosses between parents of proven highest genetic quality (Gill, 1983b; Zobell *et al.*, 1984).

Importations

Most conifer seed imported into Britain has been collected from wild forests or from unselected stands growing inside the natural range of the individual species. Relatively few countries outside the European Economic Community, and from which Britain wishes to

3

import seeds, have yet begun to export seed from selected stands. The main exceptions are Czechoslovakia, Romania and Poland. Almost all seed from north-west America is still derived from general collections which often come from cone caches collected by squirrels. Although the origin of such seed can now be officially certified, the seed is regarded as being inferior in genetic quality to seed of the same origin collected from selected stands in Britain for the reasons stated in the introduction to this chapter.

Various categories of seed are available from different countries and the certification schemes in operation within them are described in Chapter 4.

British sources of seed

Seed stands

Stands for seed production of *Pinus sylvestris, P. contorta, Larix* spp. and other species which begin to flower when relatively young are best selected in healthy crops around the time of canopy closure. Repeated heavy thinnings are required to isolate the best individual trees at this stage and can be made where there is no serious risk of windthrow to the remaining crop. The thinnings must include the less vigorous trees and those with poor stem and crown form (which would otherwise serve as undesirable pollinators) to promote the development of long, well lit, flower-bearing crowns on the remaining seed trees and to provide easy access for cone collecting teams and their equipment. Thereafter such stands are managed and maintained permanently on the National Register of Seed Stands until finally felled, preferably when bearing a very heavy cone crop. This may not coincide with the normal rotation age. Other conifer species such as *Pinus nigra* var. *maritima, Picea sitchensis, P. abies* and *Pseudotsuga menziesii* are best left to develop as commercial timber crops before registration.

In good seed years most flowering usually occurs in older crops on the upper crowns of the dominant trees. Consequently the risk of heavy pollen contamination of the dominants from less vigorous components of the crop is reduced. The first year such stands show heavy flowering they should be recommended for inspection and possible registration (see Chapter 5). Selected dominant trees bearing good cone crops should be felled before seed fall during the period when cones are ripe or nearly so. Wherever possible the complete felling of very high-quality, mature stands should be deferred until a good seed year occurs. Within any one year and any one region of provenance (see p.27) stands of one species and origin may be treated as one entry in the National Register and cones and seeds therefore may be mixed and handled as one.

Seed stands of *Betula* spp. are managed in much the same way as *Pinus contorta* except that seed collections are normally made from fellings before the developing seed catkins begin to disintegrate in August. Seed stands of *Quercus* spp. and *Fagus sylvatica* are selected from older crops and, because their mast is harvested from the forest floor, management treatments are concerned mainly with the removal of inferior trees and the control of ground vegetation to facilitate easy collection.

Where there is a need to make special seed collections for conservation purposes, for example for the renewal of certain native Scots pinewoods, it is essential that the parent population be adequately sampled to ensure that the seed reasonably represents the parent population. When planning such work, advice on how to sample the crop can be obtained from the Forestry Authority's Research Division.

The processes involved in the initiation of flowers are fully described in Chapter 6. Increased flowering in seed stands often occurs in a year following a prolonged, warm, dry and early summer. Flowering can sometimes be artificially enhanced by making two slightly overlapping half-circular cuts, 5-10 cm apart and on opposite sides of the bole some 15 cm above ground level (Figure 2.1). The cuts which should pass through the bark and into the wood to a depth of 1/2-1 cm are made in February-March using a chain saw. Trees which fail to

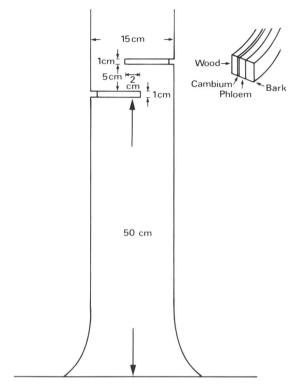

Figure 2.1 Dimensions and locations of chain saw cuts for partial ring-barking of mature trees to stimulate flowering.

respond to this treatment in the following year usually do so in either the second or third years thereafter. Applications of fertilisers to stimulate flowering in seed stands have seldom been effective and are no longer recommended. Flowering also tends to follow other periods of stress, for example, when roots are loosened by and broken during winter gales.

Seed orchards

Seed orchards are plantations of selected, vegetatively propagated trees (clones) or seedlings of known identity, isolated to reduce pollination from inferior outside sources and intensively managed to produce frequent, abundant, and easily harvested seed crops. Orchards composed of clones or seedlings of known parentage which have not been progeny tested are termed not-tested orchards (designated NT in British identity numbers).

Those composed of clones or seedlings with well-matched flowering times and proven to have outstanding genetic qualities are termed tested orchards. In order to market seed from an orchard of any of the species covered by the Forest Reproductive Material Regulations, all the component clones must have been derived from parents growing in the same region of provenance in Britain, although this will be changed if, as is likely, the EEC Directives are so amended in the current revision (1991).

Grafted trees which are normally used in clonal orchards are very expensive to produce and establish but with an initial spacing of 6 m × 6 m only require 550 plants per hectare. Relatively young grafted plants often flower and collectable seed crops in *Pinus sylvestris*, *Larix* spp. and *Picea sitchensis* often occur from the seventh year after planting. The main limitation to commercial seed production in earlier years is often insufficient pollen to ensure a high set of full seed.

Seedling orchards, which are composed of seedling plants of known pedigree, are only suited to species such as *Pinus contorta* and *Betula pendula* which flower at an early age. The task of creating seed for the initial planting stock involves artificial pollination methods and requires time, a good supply of suitable flowering material and specialised knowledge and skills; the cost of producing planting stock for seedling orchards is, therefore, very high.

The location, design, establishment and maintenance of an orchard demands a high level of expertise and a knowledge of the flowering habits, genetic quality and compatibility of the component clones or families. Fuller details were given in Forestry Commission Bulletin 54 *Seed orchards*, now out-of-print, but anyone wishing to establish a seed orchard is advised to first consult the Head of the Forestry Authority's Tree Improvement Branch.

The relative value of different seed sources

The improvement in growth rate expected from the commonest types of seed sources is indicated in Table 2.1. For the sake of comparison the expected improvement to be gained from using material of the highest genetic quality derived from vegetative reproduction is also given.

For any given species, improvement by selection and breeding is affected by:

- the genetic variability — usually indicated by the extent of the natural range of the species and the amount of between- and within-regional variation;
- the intensity of both natural and artificial selection pressures;
- and the heritability (the portion of variation which is genetic, as opposed to environmental) of the character to be improved by the selector, for example, stem straightness and branch angle.

Heritability estimates can only be determined from progeny and/or clonal tests.

For obvious reasons the calculated gain cannot be achieved unless the crop is raised under the best silvicultural practices.

REFERENCES

GILL, J.G.S. (1983a). Genetic improvement in some forestry practices — with special reference to natural regeneration. *Scottish Forestry* **37** (4), 250–258.

GILL, J.G.S. (1983b). Comparisons of production costs and genetic benefits of transplants and rooted cuttings of *Picea sitchensis. Forestry* **56** (1), 61–73.

ZOBELL, B.J. and TALBERT, J.T. (1984). *Applied forest tree improvement* (Chapter 14 — Gain and economics of tree improvement), 438–458. John Wiley and Sons, New York.

Table 2.1 Types of seed and plant source and indications of expected genetic improvement

Types of source and method of selection	Improvement
A. Seed	
Mass selection	
1. Unselected – taken throughout natural range of the species.	Datum. No gain.
2. Young stands derived from 1 and planted throughout the country or region after some natural and artificial selection. (Theoretical).	Very slight positive gains; or slight loss if precocious fruiting correlated with below-average timber yields.
3. As 2. but collections from felled mature 'home' stands previously thinned to favour best dominants.	Slight gain over 1.
4. Proven best origins (determined from tests) within the native population.	Considerable gain over 1.
5. Proven best provenances on forest sites within regions in the country of introduction.	Slight gain over 4. Gain will vary according to precocity of flowering, age of source and type and intensity of previous thinning.
6. Randomly thinned seedling seed stands based on material derived from 4 or 5.	Similar to 4 or 5.
7. Selected superior trees in superior stands of proven best origins within the native populations.	Slight gain over 4.
8. As 7 but collections from 'home sources'.	Slight gain over 4 and 7.
9. Randomly thinned seedling seed stands as in 6 but derived from 7 and 8.	Similar to 7 or 8.
10. *Untested* highly selected clones or seedling families in seed orchards.	Slight gain over 7 or 8.

Genotypic selection

11. Orchards thinned to leave proven best 10–20 clones.	Considerable gain over 1.
12. Clonal or seedling orchards composed either of highly selected proven clones, or families known to have high general and/or specific combining abilities.	Very considerable gains over 11.
13. From second and subsequent generation orchards.	Modest to high gain over 12 – generally decreasing in 2nd and subsequent generations.
14. Inter- or intra-specific hybrid orchards in which components are inter-fertile.	Variable. Most hybrids have characters intermediate between the two parents. Where hybrid vigour is obtained improvement can be exceptional in the F_1 generation.

B. *Vegetatively propagated plants*

Mass selection

1. Clones selected from untested individual trees of best origin.	Similar to A9.
2. Clones selected from untested individual trees growing in the best families generated from uncontrolled crosses in the breeding programme.	Modest gain over A11.

Genotypic selection

4. As 1 but from trees tested and approved in clonal tests.	Considerable gain over B2.
5. As 4 but from selected *tested* and approved trees in clonal tests.	Highest gains of all.

Notes:
Slight gain	=	2–3% improvement in rate of growth
Modest gain	=	3–5% improvement in rate of growth
Considerable gain	=	10–20% improvement in rate of growth
Very considerable gain	=	over 20% improvement in rate of growth

For definitions of terms used in this table the reader should consult the Glossary.

Chapter 3
The Choice of Seed Origin by Species

by **R. Lines**

Introduction

In general it is preferable to use seed from British stands or from seed orchards based on trees which have been satisfactorily tested in Britain. There are, however, no suitable home sources of some species and even where there are such sources, difficulties in collections do arise. In these cases it is necessary to import seed. Imported seed may also be preferred when the identity of the home origin is doubtful. It would be unwise to collect from such stands just because the cost is low.

The same four principles apply to most species when considering the choice of species (Lines, 1965).

1. There should be a very broad agreement between the site factors at the seed origin and at its planting site. The upland climate in Britain which, according to Wood (1974) is 'bad oceanic' rather than Continental/ Alpine, is very favourable for tree growth. Seed from trees grown in climates as different as the Alaskan coast at 61°N and the high elevation Californian Sierras at 38°N will grow well. Seed originating from areas with a highly Continental climate, such as *Larix decidua* from the French Alps, will suffer badly from spring frosts.

   Differences in day length will also affect species. *Picea sitchensis* from low latitudes in California, for example, may be killed by autumn frosts, whereas the same species from high latitudes in Alaska will set their buds so early in summer that shoot growth is limited. Low winter temperatures may cause foliage browning in some southern seed sources, some of which may even die in exceptionally cold winters. The main factors to be considered when choosing a seed origin are, however, the resistance to spring and autumn frosts.

2. There should be some knowledge of the genetic, morphological and physiological variation throughout the natural range of the species. Some species such as *Pinus contorta* have a great morphological variation, whereas *Picea sitchensis* is morphologically very similar throughout the whole of its 1800 mile range. *Picea sitchensis* does, however, vary greatly in its periodicity of growth.

3. The way in which the present distribution has been affected by events during and after the Pleistocene period should be considered. The effects of the last Glacial period and the subsequent movement of the populations have an important influence on seed selection. From example, *Picea abies* re-invaded Scandinavia via Lapland and only that part of the population whose genes enabled it to survive was able to spread west and south (Lindquist, 1947). Accordingly seed sources from much further south in Sweden grow more slowly than those from Estonia.

4. A decision has to be made on the the main objective of establishing the trees. Within a species several options may exist. A different choice may be possible depending on whether top priority is given to maximising growth, getting maximum survival, having resistance to pests or diseases, or obtaining a high wood quality.

Because of the history of provenance experiments in Britain (Wood, 1974), there are wide variations between species in the reliability of the information available on seed sources. *Pinus contorta,* for example, has been tested in about 80 trials with some 330 seed lots over a period of nearly 40 years, from the 1930s to the 1970s. Many plots are large enough to give long-term yield data. The terpene method (Forrest, 1977) may also help in identifying unknown seed origins of this species. By contrast almost nothing was known about the variation within *Abies grandis* until the 1970s and the first comprehensive experiments were only planted in 1978/79.

For many species the best choice of seed origin is clear. With other species, advice can be given on which sources to avoid but optimum choice is not yet known. Further information and distribution maps for each species are given in Forestry Commission Bulletin 66 *Choice of seed origins for the main forest species in Britain* (Lines, 1987a).

Conifers

Picea sitchensis (Sitka spruce)

Provenance trials cover both the natural range and the range of British sites adequately. There are 47 experiments with 185 seed lots. The main series were planted in 1960 and 1974 so that long-term yield data are not yet available.

As might be expected with a species which is restricted for the most part to a narrow coastal zone, the main pattern is a clinal variation with latitude in response to photoperiod and the length and warmth of the growing season. Three factors complicate this simple pattern.

1. The most southerly origins set their buds so late in the season that in many nurseries there is a strong risk of autumn frost. This applies particularly to those from Oregon and California, though in some seasons any origin below about 50°N may be damaged (Kraus and Lines, 1976).

2. There is increasing evidence that some of the sources from the Queen Charlotte Islands, BC, are more vigorous than their latitudinal position would suggest in a simple cline for increasing vigour in a southerly direction. This was less apparent in the 1960 series of experiments which contained only one seed lot from the Queen Charlotte Islands. Later experiments in 1968 and 1974 with several seed lots have shown that on northern sites this group of seed origins is growing as fast or even faster than those from Washington or Oregon. This could be a result of the survival of vigorous interglacial populations on island refugia when the mainland populations were eliminated (Wood, 1955).

3. Based on results at 10 years from the latest series of comprehensive experiments a clear site × origin interaction is developing, with the southern origins growing best in south-west England and Wales, whereas in the north of England and southern Scotland these origins grow comparatively less well (Lines, 1987b). In the north of Scotland those from Queen Charlotte Islands are usually among the fastest growing and even Alaskan origins perform quite well, compared with their very slow growth on the most southerly sites. Phenological studies do not show significant differences in flushing-date of different seed origins, so that there is little hope for selection against damage from spring frosts. It should be possible to select very late flushing individuals and propagate these vegetatively. There is a very wide range in time of bud-set and here it is possible to select origins with rapid growth rate and early bud-set (Kraus and Lines, 1976).

The choice of seed origins of *Picea sitchensis* for British conditions is summarised in Table 3.1. However, it should be noted that on the most fertile sites, trees from the fastest-growing origins may grow so rapidly that a proportion will have such wide rings that timber strength might be unacceptably low.

Table 3.1 Summary of seed origin choices of *P. sitchensis* for very exposed and average sites in Britain

Location	Very exposed	Average exposure
England and Wales below 53°30′N	Q.C.I. Vancouver Island W. Coast Washington N. Oregon	W. Coast Washington
N. England and S. Scotland below 56°N	Q.C.I. Vancouver Island	Q.C.I. W. Coast Washington N. Oregon
Northern and W. Scotland	Q.C.I.	Q.C.I. Vancouver Island W. Coast Washington

Note: Preferred choice is listed first.

Picea abies (Norway spruce)

Provenance trials began in 1928, with 90 seed lots tested before 1953, including 22 in the 1938 IUFRO trial. Forty-seven seed lots were included in the 1965–71 experiments and in 1968 the new IUFRO trial included 1100 seed lots. Many of the older trials are unsatisfactory due to poor design or losses due to fire and frost.

It has become clear from the results of the more recent experiments that many of the older plantations in Britain are derived from poor seed origins. As this species does not cone abundantly before 30 years of age, it will be many years before the stands of superior seed origins planted recently can be used for home collection.

For forestry, the desirable characters required in *P. abies* are rapid growth, good stem form and late flushing. The latter is strongly associated with seed origin and the geographical pattern of flushing shows up best in the comprehensive 1968 IUFRO trial (see Figure 1 in Lines, 1973). Very late-flushing origins come from north-east Poland, south Latvia, White Russia and Romania. The latest results from these experiments after 11 years show that height growth was best with seed origins from a band stretching from Czechoslovakia through to south Poland and the Romanian Carpathians; also in the Harz Mountains and north-east Poland. Some individual Austrian origins also grow very well but flush much earlier (Lines, 1979a). The results from these 1968 IUFRO experiments are in broad agreement with long-term results from the 1938 IUFRO experiment.

Unlike *P. sitchensis*, there is little evidence for provenance × site interaction with Norway spruce in Britain. Seed origins combining late-flushing and rapid growth are: Czechoslovakia, north-east and south Poland, and Romania.

More seed is used for Christmas tree production than in normal forest planting; for this market, crown shape is more important than rapid growth, which may result in lanky plants. Some southern German sources appear to be most suitable for this purpose but may be prone to damage from late-spring frosts.

Pinus sylvestris (Scots pine)

There should normally be adequate supplies of superior quality home-collected seed to satisfy requirements. There are 28 provenance experiments containing about 150 seed lots. The older experiments were reviewed by Lines and Mitchell (1965). Despite difficulties associated with experimental design, the overall results are clear. In the 23 Scottish experiments, few Continental origins compare favourably with Scottish sources and there is little difference in height between those from the ancient Caledonian forests and those of unknown planted origin. In the southern and more Continental climate of Thetford Forest, seed lots from the French/German border and the North German Plain outgrew Scottish provenances. Seed lots from the extremes of the range in northern Scandinavia, southern Spain and Caucasus all failed.

In general, seed should be returned to its place of origin when grown in Britain. Thus Scottish origins should be grown in Scotland and English provenances grown in England but not much reduction in rate of growth will result if the sources are exchanged.

It will rarely be necessary to import seed but if it is, then for southern and eastern parts of England selected EEC seed sources in the North Vosges Mountains of France and from West Germany at a latitude of 50°–52°N should be suitable. For northern England and Scotland, EEC seed sources from north Germany or selected stands in Poland could be used safely. Because native Scottish sources have very blue-green needles, they offer good prospects for export of seed to North America for the Christmas tree trade.

Pinus contorta (lodgepole pine)

This species is highly variable (Critchfield, 1980) and much effort has been required to discover the most suitable seed origins. Fortunately it commences flowering at an early age, and as the establishment of seedling seed stands has continued in parallel with provenance trials, the results from these can be utilised at an early stage. As British stands flower much more prolifically than some of the stands in N. America (such as coastal muskegs) cone collection costs are much lower. Seed imports should therefore soon become unnecessary provided adequate isloation of different origins has prevailed. In the long-term, selected hybrids combining good form and rapid growth from different groups of origins are likely to supplant simple choices of seed origin.

There are about 80 provenance trials in Britain with some 330 seed lots covering nearly the whole of its enormous geographical range. However, the mass of collections come from the coastal strip (Alaska to California) and, in the interior, from the Skeena River to the Southern Interior of British Columbia. The area to the east of the Rocky Mountains in Canada and to the east of the Cascade/Sierra Mountains in USA is not well represented, as these sources were soon found to grow poorly here. The earliest experiments were planted in 1928 and the latest comprehensive ones in 1972 (Lines, 1976). Aldhous (1976) described the characteristics of the seed zones in north-west America in relation to the use of *P. contorta*

pine in Britain. The main reports on seed origin variation in Britain are by Macdonald (1954) and Lines (1966, 1976, 1980a).

The main features of each regional group of seed origins are now becoming known and it is clear that no single source possesses all the desirable characters. The experiments planted in the 1970s have given more information about certain coastal origins which have characteristics intermediate between the extreme north and south coastal forms. The characteristics of the various seed regions are shown in Table 3.2. Before making his choice of origin the forest manager must have a good knowledge of the range of site conditions to be planted and must also rank his priorities among the tree characteristics. For example, if maximum growth vigour is ranked first, then he may have to accept the risk of early instability and coarse form. If superior stem form is chosen, it is likely to be accompanied by lower volume per hectare, a higher fertiliser requirement and possibly a lower resistance to severe exposure.

Provenance × site interaction is fairly strong in relation to latitude, exposure and fertility. There is also some evidence that seed lots from the drier, very Continental climates are less healthy in high rainfall areas of western Britain, though sources from heavy rainfall sites in North America appear healthy on the driest sites. In the north of Scotland not only do the northerly seed origins perform better than they do in southern Britain, but the southern coastal sources perform less well and those from California may not survive at all. In upland sites as far south as Devon all the main seed sources can be used; on lowland sites below latitude 54°N, the risk of serious attack by the pine shoot moth *Rhyacionia* restricts the use of lodgepole pine. On the most severely exposed sites only coastal origins from Alaska, Queen Charlotte Islands, the west coast of Vancouver Island and some Washington sources are recommended. The origins most susceptible to exposure are those from Lulu Island, the south east of Vancouver Island and the Puget Sound. In general, the coastal group of seed sources are capable of growing with a

Table 3.2 Choice of main seed sources – lodgepole pine

Seed source	Growth rate	Crown density	Stability	Stem form	Tolerance of exposure	Tolerance of infertile soils	Notes
Coastal							
Alaska	Low	V.high	Good	Good	V.high	High	Grow best in N.Scotland Home sources available.
N.Coastal Includes QCI and N. and W. Vancouver Island	Moderate	V.high	Good	Good	High	High	Hardy general purpose seed sources. Seed supply difficult
Vancouver S. and E. Vancouver Island and adjacent mainland	High	High	Poor	Moderate	Low	Moderate	Avoid exposed sites.
Puget Sound Rain shadow of Olympic Mts.	V.high	High	Poor	Moderate	Low	Moderate	Avoid exposed sites.
Washington Pacific Coast	V.high	V.high	Poor	Poor	V.high	High	Use of tubed seedlings can improve stability but not snow damage.
Oregon	V.high	V.high	Poor	Poor	V.high	High	Avoid sites in N. Scotland.
Inland							
Skeena River	High	Moderate	Good	Good	Moderate	Moderate	Good general purpose sources for less exposed sites and for mineral soils.
Bulkley River	Moderate	Low	Good	Good	High	Low	Second choice to Skeena. Stem form on some sources superior.
Central Interior BC	Low	Low	Good	Good	High	Low	Hardy, well proven. Good stem form.
Southern Interior BC Includes N. Washington and Idaho	High	Low	Moderate	Moderate	Moderate	Low	Variable within this group. Best are a good choice on many sites.
East of Rockies Alberta, Saskatchewan, Montana	V.low	Low	Good	Good	Moderate	Low	Usually another source will be preferable.
Cascade Mountains of Washington and Oregon	V.low	Low	Good	Good	Moderate	Low	Good form, but very slow-growing.

lower nutrient input than inland sources. The latter quickly show potassium deficiency on poor sites, particularly oligotrophic peats and may require both phosphate and potash at planting.

Where *Pinus contorta* is used in nutritional mixtures with Sitka spruce on poor fertility sites (McIntosh, 1983), it is important to use slow-growing Alaskan origins.

There is a complex interaction between tree stability, site factors and seed source (Lines, 1980b). Fast early growth (of any seed origin) can lead to basal sweep. As the incidence cannot be easily controlled by cultural measures, choice of an inherently slower-growing origin is a more certain remedy.

Pinus nigra var. *maritima* (Corsican pine)

The European black pine *Pinus nigra* has a discontinuous range in the mountains of Central Europe and in the northern Mediterranean region from Spain to Turkey. It shows considerable morphological and physiological variation and Vidakovic (1974) has suggested that as there are various transitional forms between the populations which some authors regard as separate species, it is best to regard these as varieties or sub-species. In Britain seed origins from a large part of the range have been tested in provenance trials, those from Corsica, Calabria (southern Italy) and Austria being of most interest.

Fifteen experiments have been planted with about 70 seed lots, though these include some from British stands and others from plantations from north of the native range in France, Belgium and Denmark. The Belgian 'Pin de Koekelare' (Gathy, 1956) and the stands at Les Barres, France (Macdonald, 1955) came originally from Calabria, while seed from Denmark originated in Corsica, Calabria and Austria. In general, no seed source has shown a marked superiority over the 'standard' sources from Corsica, though there is evidence that some stands which have grown well in northern countries (including Britain) may

give slightly superior progeny than direct imports from Corsica. This species grows best on warm, sunny sites in the south of Britain. On elevated sites with a cool, wet climate it is liable to attack by *Brunchorstia pinea (Gremmeniella abietina)* shoot die-back. Read (1967) considered that under unfavourable conditions "all provenances appear to be equally susceptible to attack". This has now been confirmed in the higher elevation experiment planted in the Pennines in 1962, whereas three lower elevation experiments there have not yet been affected by *Brunchorstia* (Lines, 1985b). Needle retention and diameter growth appear to be better on the second generation Calabrian sources than either the first or second generation Corsican ones. The Calabrian pines tend to have heavier branches than those from Corsica.

For dry, sunny sites in the south of England and the Midlands, home sources of Corsican origin are recommended. If home seed is unavailable, then seed from registered EEC stands in Corsica region 01(2A) or 01(2B) are the next choice. For northern, cool or high-rainfall sites, the risk of *Brunchorstia* die-back is very real and seed collected from second generation stands of Calabrian origin, e.g. Koekelare, Les Barres or from healthy stands in northern Britain offer the best hope for success.

On thin soils over chalk or limestone the Austrian variety is more healthy than the one from Corsica, which is normally found on acidic granite soils. The former may also be preferred for severe coastal exposure.

Larix decidua (European larch)

European larch has a discontinuous natural distribution occurring in four main areas: a. Alpine, b. Sudeten Mountains, c. Polish, d. Tatra Mountains (MacDonald *et al.*, 1957). However, this species was planted widely outside its native range in many European countries during the 18th and 19th centuries. Alpine seed was most widely available commercially and large quantities were used throughout Europe, often successfully, though sometimes with disastrous results in terms of canker and die-back.

The distribution of the natural stands of Sudeten, Polish and Tatra larches is much more restricted. As with Alpine larch, seed was taken from these natural Carpathian stands and used outside their natural range. Many excellent stands resulted in Czechoslovakia, Germany and Silesia (that part of pre-war Germany which is now in south-west Poland).

Confusion has arisen because there has been a tendency among foresters and particularly seed merchants to label seed from anywhere in Czechoslovakia, south-west Poland and the Tatra Mountains as 'Sudeten larch', and to fail to distinguish between seed from local native sources and seed from plantations the seed source of which has often not been known.

Provenance variation has been studied for many years on the Continent and in Britain (Edwards, 1954, 1962; Lines, 1967). Choice of provenance cannot be considered in isolation from the risk of canker and die-back (Pawsey and Young, 1969; Buczacki, 1973). The latest position is summed up by Lines and Gordon (1980). There are 13 southern experiments and 26 northern ones with over 200 seed lots (including home collections). No new experiments have been planted for 21 years, so that there are no tests which include origins from registered seed stands in EEC countries and few containing authentic Sudeten larch from the Jesenicky Mountains of Czechoslovakia.

Most trials have shown an important interaction between seed origin and site factors as reflected in their growth rate and the incidence of canker and die-back. In more favourable sites in England, as well as some in Scotland, it was soon apparent that the choice of seed origin was not critical for good survival and reasonably healthy growth. On less favourable sites in Scotland and in Wales high elevation Alpine seed proved highly susceptible to die-back and at times resulted in total failure before the stand was 20 years old. Even the more resistant Scottish and Silesian provenances were quite badly damaged on the worst sites. Only *Larix leptolepis* or *L. × eurolepis* proved safe to grow on these die-back susceptible sites. Despite considerable research efforts, much remains to be found out about the complex of canker and die-back, with its associated problems of frost and *Adelges laricis* infestations.

The four most desirable silvicultural characteristics for *L. decidua* grown in Britain are: high growth-rate, good stem-form, good resistance to canker and die-back, and late flushing. None of the origins so far tested has shown all these characteristics. In the absence of inter-provenance hybrids and with the continuing shortage of *L. × eurolepis* seed, truly satisfactory sources of seed are impossible to find. The choice of origin must therefore be influenced by the site to be planted.

Table 3.3 illustrates, by ranking, the main features of the group of seed origins tested in Britain. The best all-round choice would seem to be the true Sudeten larch from within the natural distribution in the Jesenicky Mountains in northern Moravia. Other sources

Table 3.3 The main characteristics of the groups of European larch seed origins tested in Britain

Origin	Growth rate	Stem form	Resistance to canker and die-back	Flushing date
Polish	1*	5	1*	4
Sudeten/Silesian/Tatra	2	3/4	1	4
'Scottish'	3	2	2	3
'N. German Lowlands'	3	2	2	3
Eastern Alps (low elevation)	3	1*	3	3
Eastern Alps (high elevation)	4	2	4	2
Western Alps (high elevation)	5	3	5	1[+]

*=best; [+]=earliest

of seed from Czechoslovakia should only be imported if some additional confirmation of their Sudeten origin can be obtained from the seed certification authorities. A satisfactory alternative would be seed from Polish western Silesia on the northern side of the Riesengebirge, but only if there is some official guarantee of its source. So little is known about Tatra origins that no conclusions can be drawn. The next choices in order would be: older registered British stands, older registered EEC stands, particularly in the German lowlands with Schlitz as first choice, and finally low elevation sources from Lower Austria. All other Alpine seed sources should be avoided.

Larix leptolepis (Japanese larch)

The natural range in the sub-alpine regions of central Honshu, Japan, extends over an area about the size of Wales, varying from 900 to 2700 m in elevation. It is divided into about 10 separate areas each with genetically isolated populations. The climate within its range is hot and damp in summer (rainfall 1200–3600 mm a year) and cool and dry in winter; it cannot be matched anywhere in Britain. Since this species comes from the latitude of Sicily it is adapted to a very different photoperiod. It is a pioneer species on recent volcanic soils. The sinuous stem form so common in Britain is very unusual in Japan. There are now extensive plantations in the northern island of Hokkaido, 8° north of its native range.

In Britain there are 10 provenance experiments containing about 40 seed lots. Pre-war experiments did not show large differences. Experiments with up to 25 seed origins planted in 1959 showed that high elevation sources and those from outlying stands grew poorly, while all the tallest ones came from an elevation band of 1680–1830 m (Lines and Mitchell, 1969). Assessments after 15 and 20 years confirm these results, with slight changes in rank order. A commercial seed lot from Hokkaido was as tall as the best native origins.

The best sources are likely to be from selected British plantations, as these will have been thinned to remove individuals with bad stem and crown features. Imported seed should come from Nagano Prefecture, preferably from the Suwa region and from an elevation of about 1700 m. If seed from this area is not available then other parts of Nagano Prefecture at the same elevation or from Hokkaido are acceptable alternatives. It is probably best to avoid buying seed from below 1300 m or from above 2000 m in Nagano Prefecture. Seed collected from British plantations may have a proportion of natural hybrids with European larch.

Pseudotsuga menziesii (Douglas fir)

Douglas fir has a very wide range in western North America, from 55°N in British Columbia to 25°N in northern Mexico. It is found from sea level up to 900 m at the north of its range and up to 1800 m in northern California. Botanists recognise a separate var. *glauca* on the east slopes of the Cascade Mountains and in the Rocky Mountains. Some also recognise a var. *caesia* as the type most commonly found in the interior of British Columbia. It is clear that the var. *glauca* has no place in commercial forestry in Britain, though it has some horticultural value. This species was one of the earliest introductions of forest trees from Western North America in 1826–27. Its early growth was so impressive and its timber so well known, that it was quickly taken into forest practice, particularly on private estates in Scotland.

Most seed imports before 1950 came from the lower Fraser River, British Columbia, or from Washington, with a few large consignments from 'USA' (unspecified) and from the interior of British Columbia. Comparatively small quantities were collected from home sources. Provenance trials were planted in the USA from 1912 (Munger and Morris, 1936) and in Europe from 1912 (Boiselle, 1953–54). In Britain there was some early controversy about the best sources to import (Richardson, 1905) and the earliest provenance trials at Radnor, Lael, and the Forest of Dean suffered in various ways (Wood, 1955). Eleven experiments have been planted in Scotland and northern England with 82 seed lots and 19 experiments were planted in southern England and Wales with 118 seed lots. The two main series were planted in 1953–54 with 13 to 19 seed lots from Washington and Oregon west of the Cascade Mountains and in 1970–72 with up to 44 IUFRO seed lots, the latter being the first experiments to encompass the whole range. A third series of four experiments compared eight seed lots from Scottish stands with five direct from North America.

Experience has shown that several factors have to be considered when selecting suitable seed sources of this species and that a decision based on early results can be misleading. For example, growth-rate in the nursery was high with southern origins, but these later proved less resistant to frost; the coastal group were more heavily infested with *Adelges* than those from interior British Columbia during the early thicket stage, but after 10 years of age the latter suffered much more severely from needle-cast fungi (*Rhabdocline*) than the coastal group. In some British stands very coarse branching and waviness of stems has been attributed to poor seed origin. Wood (1955) thought that "it seems more likely that certain strains have it in them to develop such characteristics on certain sites, perhaps sites much moister than they are used to".

The results of origin experiments in Britain have been reported mainly in the Forestry Commission's *Report on forest research* (see especially *Reports* for 1961, 1964, 1966, 1968 and 1970 and later papers by Lines (1980c) and Pearce (1980)). Over a wide area of Britain the same seed sources have grown well. These come from a U-shaped area comprising the lower slopes of the Cascade Mountains in northern Washington down to the southern end of the Puget Sound and up again to the foothills of the Olympic Mountains and the west side of the Olympic Peninsula. Continental foresters have highlighted the Darrington area as an optimum source. While trees from Darrington have grown well in Britain, it appears that many alternative sources are equally acceptable and Elma, Forks, Cathlamet, and Enumclaw in

Washington have all grown well. More southerly sources from Vernonia, Coquille and Waldport in Oregon have grown well on southern sites though these may be too risky in Scotland (Waldport had very poor survival at Dunkeld, Tayside). Some Vancouver Island sources grew well in Scotland and quite satisfactorily in England.

Much seed has come from dry gravelly sites around the Puget Sound. Such sites are very deficient in nitrogen compared with those on which *P. menziesii* is planted in Britain.

There was apparently little difference between the IUFRO seed sources and commercial seed lots from adjacent areas. There is sufficient variation in performance of any one seed origin on a range of sites to conclude that matching specific provenance regions in North America to specific regions in Britain is unlikely to be worthwhile. It is more important to avoid slow-growing seed sources, such as those from the northern interior of the range (e.g. Fort St. James and Williams Lake), and the most southerly sources for planting on northern sites.

Registered stands in Britain, even if of unspecified origin, are acceptable sources of seed.

Abies grandis (grand fir)

Abies grandis is potentially the most productive species in Britain (up to Yield Class 34) though it has been used on a very restrictive scale so far (3067 ha planted up to 1976). It was introduced about 1850 and used mainly in arboriculture. The natural range can be divided into three main regions of pure *A. grandis* and two where intergrades with *Abies concolor* are found (Steinhoff, 1980). These are: a) coastal lowlands of British Columbia, Washington, Oregon and California, including lower elevations on the west slopes of the Cascades; b) eastern slopes and higher elevations in the Cascades north of about 45°N, and c) northern Idaho and the south-east of British Columbia. In these pure *A. grandis* occurs. In regions d) Klamath Mountains and the Cascades of south-west Oregon and e) north-east Oregon and west central Idaho, various intergrades are found which are difficult to ascribe with certainty to either *A. grandis* or *A. concolor*. *A. grandis* grows from sea level on Vancouver Island to 2000 m in the interior, and under climates varying from near Mediterranean in California to an Alpine type on the upper slopes of the eastern Cascades. Up to 1968 seed was imported from all parts of this range; 28 per cent came from British Columbia and west of the Cascades, while 54 per cent came from east of the Cascades, the remainder being of unspecified 'USA' origin. As the most easily collectable cone crops lie east of the Cascades, the chances are that much of this seed came from that region.

There were no provenance trials with this species in Britain before 1967–68, when three experiments with five seed lots were planted. Two more experiments were planted in 1974 with 24 seed lots (Lines, 1979b). The IUFRO collection with 36 seed lots was sown in 1976–77 and planted in 1978–79 at 12 sites. There are thus no long-term data on growth of various seed origins, but judging from experiments in Denmark (Kjersgard, 1980) the early results should provide a good indication of growth up to 25 years. First results of British experiments show that production of different seed origins may vary from below Yield Class 12 for some sources east of the Cascade Mountains up to Yield Class 24 and above for sources west of the Cascades.

The following seed origins should be used: a. Washington, low elevation origins to the north, east and south of the Olympic Mountains, e.g. Port Angeles, Sequim, Louella, Shelton and Matlock; b. Vancouver Island, e.g. Campbell River, Metchosin, Nanaimo; c. Oregon, Coast Mountains, e.g. Coquille, Gold Beach, Port Orford. These vigorous southern sources are less hardy against autumn and winter frosts than more northerly origins. Seed from registered British sources, of suitable origins, is also acceptable.

All origins from east of the Cascade Ridge, especially in Oregon and Idaho, should be avoided. It would be wise to avoid collecting from British stands of unrecorded seed origin, as even the less vigorous sources can appear satisfactory when there is no local comparison with a superior seed origin and the stand is on a fertile site.

Abies procera (noble fir)

Abies procera is capable of high volume production on a range of sites and it can be considered as a possible alternative to Sitka spruce on high, exposed sites of moderate fertility (Aldhous and Low, 1974). It has a more restricted range than *A. grandis*, occurring mainly on the west side of the Cascade Mountains from Washington to northern California, where it merges with *A. magnifica*. *A. procera* also occurs on the higher peaks of the Coast Range. It is only rarely found below 500 m and all the larger stands are at an elevation of between 900–1500, rising to 2400 m in California (Fowells, 1965). Annual precipitation in that region exceeds 1800 mm, though much of this falls as snow, giving a growing-season rainfall of only 500 mm.

A. procera was introduced in 1825, and it became popular as a specimen tree when further consignments arrived in the 1850s. It is unusual among species of the genus *Abies* in coning when less than 25 years old, and as a result much home-collected seed has been available. Between 1920 and 1950, 76 kg of seed was imported, compared with 1290 kg collected in Britain. The North American origin of many older stands is impossible to trace.

There is virtually no information on seed source variation in this species. Trials were begun in Denmark in 1969 with seven seed lots but no results have been published. In Britain two experiments were planted in 1968 but these compared only one Oregon seed lot with seven collected in Britain or Denmark. The performance of the home seed lots was similar to that of the imported one. Wood (1957) pointed out that poor progeny were sometimes obtained from early home collections (due to self-fertilization), but this probably applies mainly to seed collected from isolated individuals and small groups. In 1983 part of the IUFRO collection of 19 seed lots was planted on five sites and these experiments will soon indicate the degree of variation which exists over a wide part of its range. It is already clear that trees from southern Oregon are showing characteristics similar to *A. magnifica* and that there is less variation in this species than in *A. grandis*.

At present no reliable advice on suitable seed sources is possible, though on general principles seed from the highest elevations and from the most southern part of its range should be avoided. Seed from good-quality stands in Britain can be considered as satisfactory as imported seed for planting in Britain.

Broadleaves

Fagus sylvatica (beech)

Fagus sylvatica is considered indigenous in the south-east of England as far west as Gloucestershire and south Wales, though it grows well on suitable sites as far north as the Dornoch Firth. Edlin (1962) noted that because of infrequent mast years, it has been regular practice for at least 200 years to import seed (and plants) in years when there is a dearth of home seed. Such seed may have come from anywhere between Scandinavia and Bulgaria. The exact origin of many British stands must therefore be open to question. Over the last 20 years there have been large imports of plants from continental nurseries, often of Romanian origin. With this species stem and crown-form is perhaps even more important than growth-rate. Brown (1953) has examined the site factors which influence form in some detail and he also looked at the effects of lammas growth and phenology, but he was unable to draw any firm conclusions about the genotypic variation due to seed source. Late-flushing seed origins are more likely to escape late spring frosts, but there tends to be a wide individual variation in flushing within a seed lot.

There are four beech provenance trials in Scotland and six in England, all planted between 1942 and 1955. They are nearly all comparisons of imported seed lots with home collections. Overall it appears that the best British seed sources tested (of uncertain origin) were equal to or slightly less vigorous than the best continental ones, with the possible exception of a uniformly fast-growing set of almost certainly indigenous provenances from the Forêt de Soignes, Belgium. Dutch seed (of unspecified provenance) was also above average. In the one experiment including Romanian origins, the latter performed about average for form and vigour. In contrast the few Danish sources tried performed poorly. Where it is possible to compare the same sources on two sites there is a tendency towards agreement in ranking, though this is not very close and seldom were the differences between origins highly significant. As regards form, the trees in these trials have a uniformly low grading, which does not suggest that any imported origin can be regarded as outstanding under the experimental conditions used, which were without any overhead cover and planting at about 1.5 m spacing.

If the supply of *F. sylvatica* seed is not limiting, the first choice should be from the Forêt de Soignes (Belgian Registered Stand No. 204401/155 P). In fact, its seed is in chronically short supply and forest managers will invariably have to accept alternative sources. Registered stands from Britain, Belgium from the Ardennes (Nassogne, Wellin) or de Gaume regions (Florenville, Arlon), France regions 02 Channel Coast, 03 Picardy, 04 NE Calcareous, 05 Northern Vosges, or Germany region 810 08 should all be acceptable sources of seed in the above approximate order of preference. Seeds from registered Danish sources should be used as a last choice.

Quercus spp. (oak)

The two native species, *Quercus robur* (English or pedunculate oak) and *Q. petraea* (sessile or durmast oak), have a wide distribution throughout Europe and reach their north-western limits in Britain, though they grow much farther north in Scandinavia. The two species have been so mixed up in planting over many centuries, that most woods contain a range of types. *Quercus* spp. have more frequent seed years than *F. sylvatica,* but there has been a long tradition of importing acorns from the continent, especially *Q. petraea* from Germany (Spessart) and *Q. robur* from France, and more recently Holland, to supplement home collections.

There are five provenance trials which include a few imported seed sources in comparison with home collections. In general the imported seed lots are no better and often inferior to the best home sources as regards height growth. However, these old experiments did not take into account the importance of seed size on later seedling growth (Jarvis, 1963) so that true seed origin effects may have been confounded. Stem form does not appear to be appreciably better in the plots of continental origin, but it should be remembered that these were unselected imports and that the progeny from registered EEC seed stands could be superior.

If available, seed from Registered British sources should be used. For imported seed the first choice should be Registered sources in Germany, regions 818 11 (Spessart), 818 03 (north-west Germany) for *Q. petraea,* 817 01 (northern Schleswig-Holstein) for *Q. robur,* 817 03 (German Lowlands) or France, Regions 01, 03, 07, 12 for both species. Seed from Registered Dutch stands should be used as a second choice as many stands in that country are in avenues where limited natural selection can take place.

Betula spp. (birch)

The two native species have not been extensively planted in Britain, so that *Betula pubescens* is more often found on wetter, peaty sites to the north and west and *Betula pendula* on drier mineral soils in the east and south. The two species are thought to hybridise where their ranges overlap (Gardiner and Pearce, 1978) and each is very variable in its leaf morphology. Latest research shows that due to difference in chromosome number, hybrids between the species are rare and are themselves sterile (Brown, 1991).

There are no replicated trials with imported seed origins. A small collection of *Betula* progenies from Norway and Sweden planted near Edinburgh in 1951 shows better stem form than progenies from Scottish collections; the former suffer more from leaf rusts. In Sweden and Finland tree improvement of these species has reached an advanced stage and it would be possible to obtain seed from improved strains, though these have been selected primarily from sources at much higher latitudes than our own and might be ill-adapted, except perhaps in Scotland.

No reliable advice can be given at present on choice of imported seed sources. On general principles, seed should not be imported from latitudes above 62°N or below 45°N nor from more continental parts of Europe.

REFERENCES

ALDHOUS, J.R. and LOW, A.J. (1974). *The potential of western red cedar, grand fir and noble fir in Britain.* Forestry Commission Bulletin 49. HMSO, London.

ALDHOUS, J.R. (1976). Lodgepole pine seed zones with reference to British requirements. In *Pinus contorta provenance studies,* ed. R. Lines, Forestry Commission Research and Development Paper 114. Forestry Commission, Edinburgh.

BOISELLE, R. (1953–54). Die Snoqualmie-Douglasie, die Douglasie der Zukunft. *Allgemeine Forst– und Jagdzeitung* **125** (2), 61–69.

BROWN, I. R. (1991). The current state of birch in Britain. In *The commercial potential of birch in Scotland.* Forestry Industry Committee of Great Britain, London.

BROWN, J.M.B. (1953). *Studies on British beechwoods.* Forestry Commission Bulletin 20. HMSO, London.

BUCZACKI, S.T. (1973). Some aspects of the relationship between growth vigour, canker, and dieback of European larch. *Forestry* **46** (1), 71–79.

CRITCHFIELD, W.B. (1980). The distribution, genetics and silvics of lodgepole pine. In *Proceedings of the IUFRO Joint Meeting of Working Parties, Vancouver, Canada 1978*, 65–94. Ministry of Forests, Province of British Columbia, Canada.

CRITCHFIELD, W.B. and LITTLE, E.L. (1966). *Geographic distribution of the pines of the world.* USDA Forest Service, Miscellaneous Publication 991.

EDLIN, H.L. (1962). A modern sylva or a discourse of forest trees. 2. Beech - *Fagus sylvatica* L. *Quarterly Journal of Forestry* **66** (3), 196–205.

EDWARDS, M.V. (1954). Scottish studies of the provenance of European larch. *Proceedings XI Congress of IUFRO, Rome, 1953*, 432–436.

EDWARDS, M.V. (1962). Choosing seed origins of European larch. *Scottish Forestry* **16** (3), 167–169.

FORREST, G.I. (1977). Geographical variation in the monoterpenes of the resin of *Pinus contorta*. In *EEC Symposium on Forest Tree Biochemistry*, EUR 5885, 55–71. Commission of the European Communities, Brussels.

FORREST, G.I. (1980). Geographical variation in the monoterpenes of *Pinus contorta* oleoresin. *Biochemical Systematics and Ecology* **8**, 343–359.

FOWELLS, H.A. (1965). *Silvics of forest trees of the United States.* Agriculture Handbook 271. USDA Forest Service.

GARDINER, A.S. and PEARCE, N.J. (1978). Leaf-shape as an indicator of introgression between *Betula pendula* and *B. pubescens*. *Transactions of the Botanical Society of Edinburgh* **43**, 91–103.

GATHY, P. (1956). Aperçu des recherches en matière de genetique forestière. *Bulletin du Société Royale Forestière Belgique* **63** (10).

JARVIS, P.G. (1963). The effects of acorn size and provenance on the growth of seedlings of sessile oak. *Quarterly Journal of Forestry* **62** (1), 11–19.

KJERSGARD, O. (1980). *Abies grandis* in Denmark: a summary. *Proceedings of the IUFRO Joint Meeting of Working Parties, Vancouver, Canada, 1978* **2**, 347–348.

KRAUS, J.F. AND LINES, R. (1976). Patterns of shoot growth, growth cessation and bud set in a nursery test of Sitka spruce provenances. *Scottish Forestry* **30** (1), 16–24.

LINDQUIST, B. (1947). The main varieties of *Picea abies* (L.) Karst in Europe. *Acta Horticulturae Bergiani* **14** (7), 249–342.

LINES, R. (1965). Provenance and the supply for forest tree seed. *Quarterly Journal of Forestry* **59** (1), 7–15.

LINES, R. (1966). Choosing the right provenance of lodgepole pine. *Scottish Forestry* **20** (2), 90–103.

LINES, R. (1967). The international larch provenance experiment in Scotland. *Proceedings of the XIV Congress of IUFRO, Munich 1967*, **III**, 755–781.

LINES, R. (1973). *Inventory provenance test with Norway spruce in Britain: first results.* Forestry Commission Research and Development Paper 99. Forestry Commission, Edinburgh.

LINES, R. (1976). *Pinus contorta* provenance experiments in Britain. In *Pinus contorta provenance studies,* ed. R. Lines, 107–109. Forestry Commission Research and Development Paper 114. Forestry Commission, Edinburgh.

LINES, R. (1979a). Results of the IUFRO 1964/68 experiments with *Picea abies* in Scotland after 11 years. In *Proceedings of IUFRO Joint Meeting of Working Parties on Norway spruce provenances and Norway spruce breeding, Bucharest 1979.* Lower Saxony Forest Research Institute, Escherode.

LINES, R. (1979b). Natural variation within and between the Silver firs. *Scottish Forestry* **33** (2), 89–101.

LINES, R. (1980a). The IUFRO experiments with *Pinus contorta* in Britain – results after six years in the forest. *Proceedings of the IUFRO Joint Meeting of Working Parties, Vancouver, Canada 1978*, 125–135. Ministry of Forests, Province of British Columbia.

LINES, R. (1980b). Stability of *Pinus contorta* in relation to wind and snow. In *Proceedings of the IUFRO Working Party on P. contorta provenance*, 209–219. Research Note 30. Institute of Forest Genetics, Swedish University of Agricultural Science, Garpenberg.

LINES, R. (1980c). The IUFRO experiments with Douglas fir in Scotland. *Proceedings of the IUFRO Joint Meeting of Working Parties, Vancouver, Canada 1978*, **1**, 297–303. Ministry of Forests, Province of British Columbia, Canada.

LINES, R. (1985a). Lodgepole pine management in the United Kingdom. In Symposium Proceedings (1984), *Lodgepole pine, the species and its management*, eds. Baumgartner, D. M., Krebill, R. G., Arnott, J. T. and Weetman, G. F., 219–229. Washington State University.

LINES, R. (1985b). *Pinus nigra* in the Pennine Hills of Northern England. *Quarterly Journal of Forestry* **79** (4), 227–233.

LINES, R. (1987a). *Choice of seed origins for the main forest species in Britain*. Forest Commission Bulletin 66. HMSO, London.

LINES, R. (1987b). Seed origin variation in Sitka spruce. *Proceedings of the Royal Society of Edinburgh* **93B**, 25–39.

LINES, R. and ALDHOUS, J.R. (1962). Provenance studies – Douglas fir. *Forestry Commission Report on Forest Research 1961*, 40. HMSO, London.

LINES, R. and GORDON, A.G. (1980). *Choosing European larch seed origins for use in Britain*. Research Information Note 57/80/SEED. Forestry Commission, Edinburgh.

LINES, R. and MITCHELL, A.F. (1965–1970). Provenance – Douglas fir. *Forestry Commission Report on Forest Research 1964*, 31; *1966*, 44; *1968*, 71–73; *1970*, 63–64. HMSO, London.

LINES, R. and MITCHELL, A.F. (1965). Results of some older Scots pine provenance experiments. *Forestry Commission Report on Forest Research 1964*, 172–194. HMSO, London.

LINES, R. and MITCHELL, A.F. (1969). Provenance – Japanese larch. *Forestry Commission Report on Forest Research 1969*, 43–44. HMSO, London.

MACDONALD, J.A.B. (1954). The place of *Pinus contorta* in British silviculture. *Forestry* **27,** 25–30.

MACDONALD, J. (1955). A tour of French forests. *Journal of the Forestry Commission* **24,** 16.

MACDONALD, J., WOOD, R.F., EDWARDS, M.V. and ALDHOUS, J.R. (1957). *Exotic forest trees in Great Britain*. Forestry Commission Bulletin 30. HMSO, London.

McINTOSH, R. (1983). Nutrition, establishment phase. *Forestry Commission Report on Forest Research 1983*, 15–16. HMSO, London.

MUNGER, T.T. and MORRIS, W.G. (1936). *Growth of Douglas fir trees of known seed source*. USDA Technical Bulletin 537. (40pp.)

PAWSEY, R.G. and YOUNG, C.W.T. (1969). A re-appraisal of canker and die-back of European larch. *Forestry* **42** (2), 143–164.

PEACE, T.R. (1948). The variation of Douglas fir in its native habitat. *Forestry* **22,** 45–61.

PEARCE, M.L. (1980). The IUFRO experiments with Douglas fir in England and Wales. *Proceedings of the IUFRO Joint Meeting of Working Parties, Vancouver, Canada 1978*, **1**, 381–388. Ministry of Forests, Province of British Columbia, Canada.

READ, D.J. (1967). *Brunchorstia die-back of Corsican pine*. Forestry Commission Forest Record 61. HMSO, London.

RICHARDSON, A.D. (1905). The Colorado variety of the Douglas fir. *Transactions of the Royal Scottish Arboricultural Society* **18,** 194–199.

STEINHOFF, R.J. (1980). Distribution, ecology, silvicultural characteristics and genetics of the *Abies grandis – Abies concolor* complex. *Proceedings of the IUFRO Joint Meeting of Working Parties, Vancouver, Canada, 1978* **1,** 123–132. Ministry of Forests, Province of British Columbia, Canada.

VIDAKOVIC, M. (1974). Genetics of European black pine (*Pinus nigra* Arn.), *Annales Forestales* **6,** 57–81. Zagreb.

WOOD, R.F. (1955). *Studies of north-west American forests in relation to silviculture in Great Britain*. Forestry Commission Bulletin 25. HMSO, London.

WOOD, R.F. (1974). *Fifty years of forestry research*. Forestry Commission Bulletin 50. HMSO, London.

Chapter 4

Systems of Seed and Plant Identification and Certification

by **A. G. Gordon** *and* **C.J.A. Samuel**

Introduction

Seed identification numbers are needed in order that particular batches of seed and the plants derived therefrom may be clearly identified and followed throughout their passage from seed stock, to nursery and finally into the forest. They have also been used as a quick means of indicating the location where the seed was collected.

The object of this chapter is to describe the various British identity systems used in the past (many such numbers are still widely found in compartment records), to give a brief summary of the systems used in countries from which Britain imports significant amounts of seed and to show how the new British identity number system is derived. Finally, a brief description of the OECD Scheme for the Control of Forest Reproductive Material moving in international trade is given.

Over the last 60 years four different systems of identifying seed lots have been widely used in Britain.

The first was operated exclusively within the Forestry Commission from 1921 to 1956. The second was a system operated by the Forestry Commission and voluntarily by some people in the private forestry sector from October 1956 to June 1973.

The next change affected seed collected in Britain and was brought about as a result of entry into the European Economic Community and the introduction of the Forest Reproductive Material Regulations on 1st July 1973 (see Chapter 5). These required regions of provenance to be defined, so as to provide a basis for identifying seed under the Regulations. The system started in October 1956 was retained until 1984 by the Forestry Commission for identifying seed imported from other countries. However, in this period other countries introduced and developed their own identification systems.

Since 1984, a system amalgamating the domestic element of the Forestry Commission system with key features of the systems used by overseas countries producing seed has been introduced by the Forestry Commission and adopted by some private British seed firms. In November 1986 this was officially endorsed by both the Forestry Commission and the Forestry Group of the Horticultural Trades Association in an effort to standardise all forest seed identification in Great Britain.

The Northern Ireland Forest Service operates its own system of seed identification and certification and is not described here.

Identification of seed in Great Britain

Period 1921–1956

During this period, individual Forestry Commission seed lots were identified by the last two numbers of the year of completion of processing, and a serial number separated by an oblique stroke. Such an identity was unique; thus no breakdown into species was necessary. Numbers were allocated in order of final processing and lists of identification numbers were published annually, circulated to all forests and printed in the *Forestry Commission Journal*. In 1965 the complete list from the years 1921–56 was accumulated in Forestry Commission Research Branch Paper 29 *Seed identification numbers* (Anon., 1965).

Early entries record identity, quantity (normally in pounds but larger hardwood seeds in bushels), species, crop year, origin and vendor. Thus:

> 22/57 6lbs; *Pinus sylvestris;* 1921 England (Military area, Camberley); own collection, extracted Bentley.
>
> 22/147 14 bushels; *Acer pseudoplatanus;* England (Thornthwaite); own collection.

The list also included seed imported from abroad (see section on imported seed lots, p.29).

Some late records provide the purity together with germination and fresh seed percentages. Some serial numbers may also have letter or combined letter and numeral suffixes:

> 55/11C3 8 lbs 15 ozs; Scots pine; 1954; Northumberland; 98.4; 81+1.

Collections made or received by Research Division were given serial numbers from 500 or from 1000 onwards by Silviculture (North) and Silviculture (South) Branches respectively.

The 1921–1956 numbering system only gives information on seed lots when used in conjunction with the published descriptions. The system which replaced it in 1956 incorporates into the identity number certain basic coded information about the source of the material.

The Forestry Commission system from October 1956 to June 1973

The permanent seed identification code introduced in 1956 is described by Matthews in the *Journal of the Forestry Commission,* 1958 (also reproduced in Research Branch Paper 29). The basis of the system was the Universal Decimal Classification (UDC) system for numerically encoding geographical and political areas of the world. Under the strict

UDC system, these codes must be enclosed in brackets and may be up to five digits long, a decimal point appearing after the first three. Increasing detail about the location is given by the inclusion of more digits in the code.

The UDC system was modified for seed identity numbers (Table 4.1). The decimal point was not used and a maximum of four digits appears in the brackets. Some other departures from the UDC system have also been made, notably in the allocation of codes to the pre-1974 British counties. These are fully described in the Research Branch Paper 29.

The numbers were built up systematically for continent, country, region and county thus:

> (4)   Europe
>
> (41)   Scotland
>
> (414)   SE Scotland
>
> (4144) East Lothian

Figure 4.1 illustrates the numbers given to Britain. Similar codes were used for foreign countries (see below). The bracketed part of the identity number was preceded by a two figure crop year. As more than one species could be collected in the same location in the same year, the crop year was usually preceded by the Forestry Commission species code letters. (If the identity number appears along with the full species name on an official certificate it was not necessary to repeat the species abbreviation.)

Examples:

> BE 66(4238) was beech (*Fagus sylvatica*) seed collected in Gloucestershire in 1966;
>
> JL 66(4238) was Japanese larch (*Larix leptolepis*) also collected in Gloucestershire in 1966.

Collections from individual registered stands (see later) were indicated and identified by a one or two digit number following the brackets; three digit numbers indicated special research collections.

> SP 66(4216)1 would have identified Scots pine (*Pinus sylvestris*) seed collected from the registered seed source in Crown Estate, Windsor during crop year 1966.

Table 4.1 The more important UDC numbers for countries/states from which seed might be imported

a. By UDC number	b. In alphabetical order by country/state	
430 Germany	Alaska	798
436 Austria	Argentina	82
437 Czechoslovakia	Australia	94
438 Poland	Austria	436
439 Hungary	Belgium	493
44 France	British Columbia	711
44945 Corsica	Bulgaria	4972
450 Italy	California	794
460 Spain	Canada	71
469 Portugal	Chile	83
47 Russia	China	510
480 Finland	Corsica	44945
481 Norway	Czechoslovakia	437
485 Sweden	Denmark	489
489 Denmark	Finland	480
492 Holland	France	44
493 Belgium	Germany	430
494 Switzerland	Greece	495
495 Greece	Hokkaido	524
4971 Yugoslavia	Holland	492
4972 Bulgaria	Honshu	521
498 Romania	Hungary	439
510 China	Idaho	796
519 Korea	India	540
520 Japan	Israel	5694
521 Honshu	Italy	450
524 Hokkaido	Japan	520
540 India	Korea	519
5491 Pakistan	New Zealand	931
560 Turkey	Norway	481
5694 Israel	Oregon	795
680 South Africa	Pakistan	5491
71 Canada	Poland	438
711 British Columbia	Portugal	469
794 California	Romania	498
795 Oregon	Russia	47
796 Idaho	South Africa	680
797 Washington	Spain	460
798 Alaska	Sweden	485
82 Argentina	Switzerland	494
83 Chile	Turkey	560
931 New Zealand	Washington	797
94 Australia	Yugoslavia	4971

Registration and identification of seed stands 1950–June 1973

From the early 1950s the Genetics Branch of the Forestry Commission Research Division was responsible for the survey and registration of stands in Britain which were suitable for seed production. Up to crop year 1972 (effectively to June 1973), stands had been placed in one of four categories A, B, C, or D:

A. seed sources suitable for management as seed stands;

B. seed sources not recommended for management as seed stands but useful for seed collection in good seed years;

Figure 4.1 Regions of provenance in Great Britain 1956–June 1973.

C. useful as seed sources only when collections made from clear fellings;

D. area of native pinewoods and young plantations of valuable origins of *Pinus contorta* or boundaries of *Larix decidua* and *L. leptolepis* where natural hybridisation might occur.

These four categories were never used as part of the seed identity numbering system. However, stands were classified as *'Plus'*, *'Almost Plus'*, or *'Normal'* according to the percentage of seed trees amongst the dominant trees in the crop. Plus and Almost Plus stands were given a permanent stand number shown as a suffix to the coded county number within brackets. Thus SP(4122)9 was the registered Plus or Almost Plus Scots pine seed stand No.9 Speymouth Forest, Morayshire.

This system was also used voluntarily by the private sector until 1973 through the auspices of the Forest Tree Seed Association, but was rendered obsolete by the changes described in the following paragraph and was modified to meet the needs of seed stand registration.

Registration and identification of seed stands in Great Britain from July 1973 onwards

Following the British entry into the European Economic Community, collection and marketing of seed of the more important forest tree species came under the Forest Reproductive Material Regulations 1973 (see Chapter 5). At the same time, the old system of location by counties in Britain was abandoned for home seed collection. To replace it, four regions of provenance were defined within Great Britain and these were numbered 10, 20, 30 and 40 – numbers which provide scope for future subdivision should the need arise, e.g. (45), (46), (47), etc. Their delineation is shown in Figure 4.2.

Within each region of provenance for each species, registered seed stands are given 4-digit numbers. The first two show the region of provenance and the second two indicate single

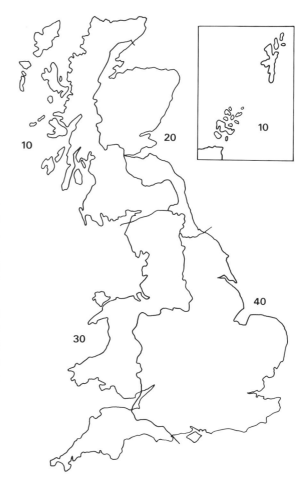

Figure 4.2 Seed Zones for Great Britain July 1973 onwards.

stands or groups of stands of the same origin and quality. Thus:

SS 84(2007): 1984 collection from Sitka spruce (*Picea sitchensis*) stand 2007 (SS registered seed stand 7 in region 20). Tay Forest District.

SP 84(2007): 1984 collection from Scots pine (*Pinus sylvestris*) stand 2007 (SP registered seed stand 7 in region 20). Glentanar Estate.

Collections from unregistered sources are identified by the region of provenance only in the brackets:

SS 84(10) A general collection in region 10 from unregistered sources.

These codes within brackets distinguish themselves from the UDC-based location codes since only those from region 40 could present ambiguity with Europe and this is ruled out by the appearance of zero as the second digit, e.g. (40) or (4017). (Stand 4100 is still many years away when chances of ambiguity will have diminished).

Unregistered seed lots which, through mixing, derive from more than one of the four regions of provenance have the word 'Britain' in the bracketed part of the ident.

SY79(Britain) Mixture of sycamore (*Acer pseudoplatanus*) collections from regions 10 and 20, etc.

Although devised for EEC species, the above system is also adopted by the Forestry Commission for seed lots of species and hybrids not covered by the Forest Reproductive Material Regulations, e.g. *Pinus contorta*, *Larix × eurolepis*, *Thuja plicata*, etc.

Native Scots pine

Seed lots from indigenous stands of *Pinus sylvestris* (Figure 4.3) have been given identities which contain the name of the source, or an abbreviation. This has appeared either in the bracketed part of the identity or as a suffix. The abbreviation 'Cal' (for 'Caledonian SP') has sometimes followed.

SP61(412)Shield	Shieldaig
SP63(Affric)Cal	Glen Affric
SP86(10)Ran	Black Wood of Rannoch

A new scheme to promote the maintenance and regeneration of native Scottish pinewoods was introduced in 1989 (Forestry Commission, 1989). This will involve a letter N for stands meeting the special conservation requirements of the scheme. The letter will be followed by a number allocated to the specific stand, e.g. SP89 (N305), i.e. seed from area 5 in native seed zone 3.

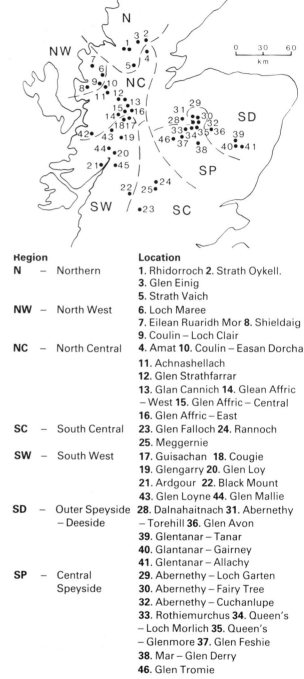

Region		Location
N	– Northern	1. Rhidorroch 2. Strath Oykell. 3. Glen Einig 5. Strath Vaich
NW	– North West	6. Loch Maree 7. Eilean Ruaridh Mor 8. Shieldaig 9. Coulin – Loch Clair
NC	– North Central	4. Amat 10. Coulin – Easan Dorcha 11. Achnashellach 12. Glen Strathfarrar 13. Glan Cannich 14. Glean Affric – West 15. Glen Affric – Central 16. Glen Affric – East
SC	– South Central	23. Glen Falloch 24. Rannoch 25. Meggernie
SW	– South West	17. Guisachan 18. Cougie 19. Glengarry 20. Glen Loy 21. Ardgour 22. Black Mount 43. Glen Loyne 44. Glen Mallie
SD	– Outer Speyside – Deeside	28. Dalnahaitnach 31. Abernethy – Torehill 36. Glen Avon 39. Glentanar – Tanar 40. Glantanar – Gairney 41. Glentanar – Allachy
SP	– Central Speyside	29. Abernethy – Loch Garten 30. Abernethy – Fairy Tree 32. Abernethy – Cuchanlupe 33. Rothiemurchus 34. Queen's – Loch Morlich 35. Queen's – Glenmore 37. Glen Feshie 38. Mar – Glen Derry 46. Glen Tromie

Figure 4.3 Indigenous Scots pine populations in native Scottish woodlands, classified into regions of biochemical similarity based on monoterpene analysis.

Seed orchards

Each seed orchard of each species has been numbered consecutively from 1 onwards. This number appears within the bracketed part of the identity prefixed by either the letters 'NT' or 'A'. NT indicates that information on the performance, in comparative tests, of component clones or families is not yet available; A indicates that the components have performed better than prescribed levels in such tests. An orchard which was planted before test information was available may pass from NT to A status on the basis of test results. This system was adopted before 1973 and is still in operation as part of the official system. It is also used by the Forestry Commission for hybrids and species not covered by Forest Reproductive Material Regulations 1977.

SP87(NT44) Seed from orchard number 44; results from progeny tests not available.

Further developments of the nomenclature

In view of the potential for commercial bulk vegetative propagation of seedlings raised from tested parent stock, a scheme for naming source material has been in use in the Forestry Commission Research Division since 1982. The general principles followed are that *tested clone or family mixtures* released to commerce are numbered serially within species, the number prefixed by an 'M'. This will appear in the bracketed part of the identity and will be preceded by a crop year in the usual way.

Details of seed stands, orchards and family mixtures are held on a register as part of the Forestry Authority Tree Improvement Branch's database. Information on all registered sources is available from the Grants and Licences Division at Forestry Commission Headquarters; a copy of the current National Register is also held at each Forestry Authority National Office.

Forestry Commission identification systems for imported seed

The systems used for identifying imported seeds followed, until 1973, the same principles adopted for identifying collections of seed from British sources.

The period 1921 to 1956

Each seed lot imported during a particular forest year was allocated a serial number at time of receipt. Its identity number comprised the last two numbers of the appropriate year of receipt followed by the serial number. Thus 23/1, 23/2 were the first two seed lots taken on stock in the forest year 1923. Imported seed lots were included in the same system as that used for British collected seeds which has been described above. Thus 23/3 and 23/4 described the third and fourth seed lots taken on stock in the forest year 1923 and could have been either imports or collections in Britain. As with the latter, these serial numbers were unique but did not in any way help to identify the location where the seed was collected. Such details, nevertheless, were published annually and were summarised in Forestry Commission Research Branch Paper 29 in 1965.

The period October 1956 to 1984

Importations during this period were identified using the modified Universal Decimal Classification System described above. The system of codes allocated to Europe and North America are shown in Figures 4.4 and 4.5.

As with British collections, identity numbers consisted of a two figure crop year followed by the country or region code in brackets. Again, as more than one species could be collected in the same location in the same year, a two letter species code was normally included. Thus:

SS71(7972) was Sitka spruce seed collected in the State of Washington, USA in 1971;

LP71(7972) was lodgepole pine seed also collected in the State of Washington, USA in 1971.

Individual collection sites or regions were indicated and identified by one or two digits as the last part of the identity number:

LP71(7972)5. 5 indicated Hoquiam, Washington;

SS71(7972)1. 1 indicated Long Beach, Washington.

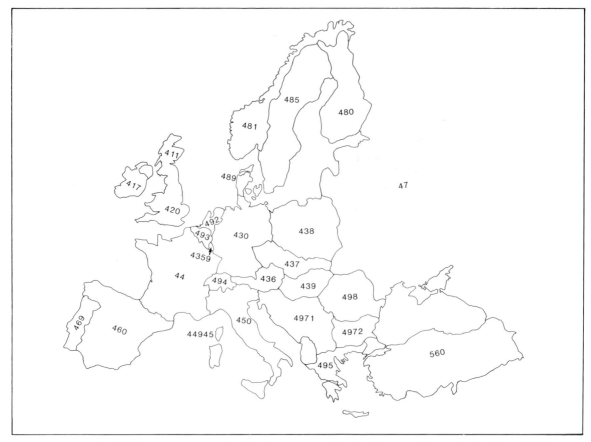

Figure 4.4 Modified UDC provenance codes for Europe used by the Forestry Commission from 1956 to date.

These last digits were kept constant for specific locations across species and years. For some imports these could be very small localities. Figure 4.6 illustrates the identity code used by the Forestry Commission for imported seed.

Despite the original intention to keep identities totally numeric in form, letter suffixes also appeared among the last digits:

(7114)13A		760ft altitude
(7114)13B	Smithers, BC	910ft altitude
(7114)13C		1070ft altitude
(7114)15	Bulkley Canyon, BC	
(7114)15A	Bear River, very close to Bulkley Canyon, BC	

SS70(7111) Lot 2 – When very large consignments were imported they were split in equal sized quantities, each handled (homogenised) as one lot. Such lots, which could have slightly different germination percentages, were called lot 1, lot 2, etc.

The letter R has commonly been used to indicate seed from a registered source in the country of origin, e.g. EL 71(4532)R was from a registered stand in Italy.

Identification of seed importations since 1984

The latest system combines part of the old UDC system with that used by the original certifying country. The identity number retains the year and the country code of the UDC within brackets but adds as a suffix the code of the region of provenance used by the original certifying country. Thus an import of 1983 Norway spruce seed from Germany Region

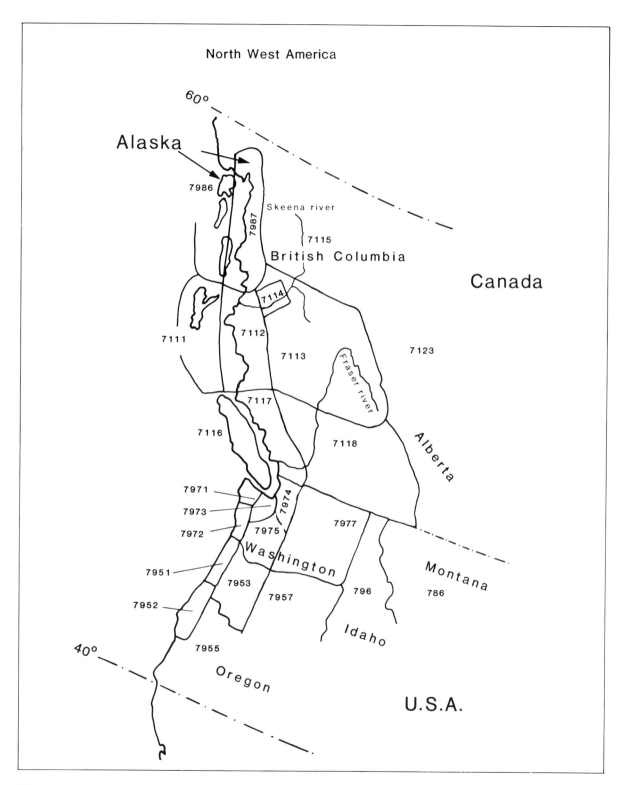

Figure 4.5 UDC provenance codes for North-west America used during the period 1956–1986.

84019 is 83(43)84019; an import of Douglas fir from Darrington (seed zone 403), Washington in 1984 is 84(797)403.

Where an EEC country certifies species additional to those covered by the EEC Directives (often called EEC species) the certification code as used by the EEC country is also used, as for the main list of EEC species.

It should be noted that more than one importation can be made from the same region of provenance in the same country in the same year, either through different seed importers or exporters, or through split seed lots. For this reason seed importers may add letters and numbers at the end of the identity number to identify themselves, the supplier and the lot, e.g. 86(492)1.1 F lot H2; F would indicate the British supplier Forestart, H the exporter and 2 the fact that since the quantity imported exceeded the maximum lot size permitted, it had to be split into two lots. Current Forestry Commission identity numbers omit both importer and exporter letters.

In order to distinguish between F_1 and F_2 hybrid larch (*Larix* × *eurolepis*), the letters E or J or both are used to designate the female parent of F_1 seed. No such suffix is used for F_2 seed.

Table 4.2 Abbreviations used by the Forestry Commission for names of conifers in commercial production in British forestry

Common name	Abbreviation
Scots pine	SP
Corsican pine	CP
Lodgepole pine	LP
Alaskan	ALP
North Coastal	NLP
South Coastal	SLP
Skeena River	KLP
C. and N. Interior	CLP
S. Interior	ILP
Sitka spruce	SS
Queen Charlotte Islands	QSS
Washington	WSS
Alaskan	ASS
Oregon	RSS
Norway spruce	NS
European larch	EL
Japanese larch	JL
Hybrid larch	HL
Douglas fir	DF
Western hemlock	WH
Western red cedar	RC
Grand fir	GF
Noble fir	NF
Lawson cypress	LC

SS

These letters denote the species – in this case Sitka spruce. Standard abbreviations are used for all the major species – see Table 4.2 for the full list. These letters are often dropped as they are redundant on Suppliers Certificates, etc., where the name in full and in Latin is written. They are still regularly used as a ready means of differentiating between different origins of lodgepole pine, eg. KLP = Skeena River, NLP = North Coastal, etc.

(7111)

These numbers inside the brackets denote the region from which the seed was collected. They are based on the UDC classification. The 1st digit records the continent, e.g. 7 represents N. America. The 2nd digit indicates the country. The 3rd digit indicates the province or other national sub-division, etc. This code tells this particular seed lot came from Queen Charlotte Islands, British Columbia, Canada.

SS 70(7111)1

70

These numbers denote the seed year in which the crop ripened. Seed years extend from 1 August–31 July and therefore this seed ripened between 1 August 1970 and 31 July 1971.

1

This digit gives information about the locality within the Queen Charlotte Islands from which the seed was collected.
Since 1985 this digit has also been used to record the region allocated to the seed lot by the original certifying country.

Figure 4.6 Illustration of use of Forestry Commission identification code for imported seed 1956–1986.

Mixing of seed

Small incoming lots of seed or those which, through issue, have been reduced to low levels are sometimes mixed with other seed from similar sources or, if testing indicates compatibility, with later larger lots from the same source. The bracketed part of the identity will be adjusted if necessary to take account of any change in region brought about by mixing and the crop year used will be that of the largest component of the mixture.

Foreign systems

Introduction

Over the last 30 years most important timber producing countries in the temperate zone have introduced their own official systems of identifying and certifying seed sources. These have normally been based upon natural distribution of the species but all vary from each other to some degree. The Organisation for Economic Co-operation and Development (OECD) was the first to publish a scheme for certificating forest reproductive material moving on international trade (see p. 47). Some countries, which had already initiated their own schemes for internal use and trade, have sought to bring theirs into line with the OECD scheme. Other countries have created new schemes which complied with the OECD scheme from the outset. With the promulgation of the Forest Reproductive Material Directive, 64/404/EEC, member countries of the EEC have been required to introduce a legally enforceable scheme using the same basic selection standards. Nevertheless, the OECD scheme remains relevant to the species not included in the EEC Directive.

Certification and identification schemes within the European Economic Community

Belgium (UDC Number 493)

The forests of Belgium are divided into two main regions of provenance called I or II which are the same for all species. These may be further subdivided for some species into regions I/1, I/2, I/3 and II/1, II/2 and II/3 (see Figure 4.7). These regions are delimited on the basis of phytogeographic, ecological and administrative criteria. The stands are selected according to the usual EEC criteria. Each stand is given a code number, an example of which is given below. Imported seed is also given a code. All codes follow seed and plants throughout their processing, storage and growth in a nursery. No seeds or plants of the EEC species can be marketed without a certificate issued by representatives of the Office National des Débouchés Agricoles et Horticoles (ONDAH) which is the Forest Authority.

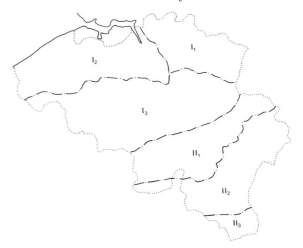

Figure 4.7 Regions of provenance for Belgium.

An example of a Belgian identity number is II/2 (a) 6132/023A. II/2 is the Belgian region of provenance, in this case the Ardennes; (a) indicates a properly managed stand in contrast to (b) an unmanaged stand; 6132 is the code given by ONDAH to the local community; 023 is the number of the selected stand in the register kept by the Département des Eaux et Forêts (and is often used on its own to identify the stand). 'A' represents the code letter for the species, here *Picea abies*. Finally the name of the seed stand is given.

For plants grown from imported seed a code such as 04 14 78/004C Darrington is used. 04 is the country code (00 Holland, 01 Britain, 02 France, 03 Germany, 04 USA, 07 Canada, etc.).

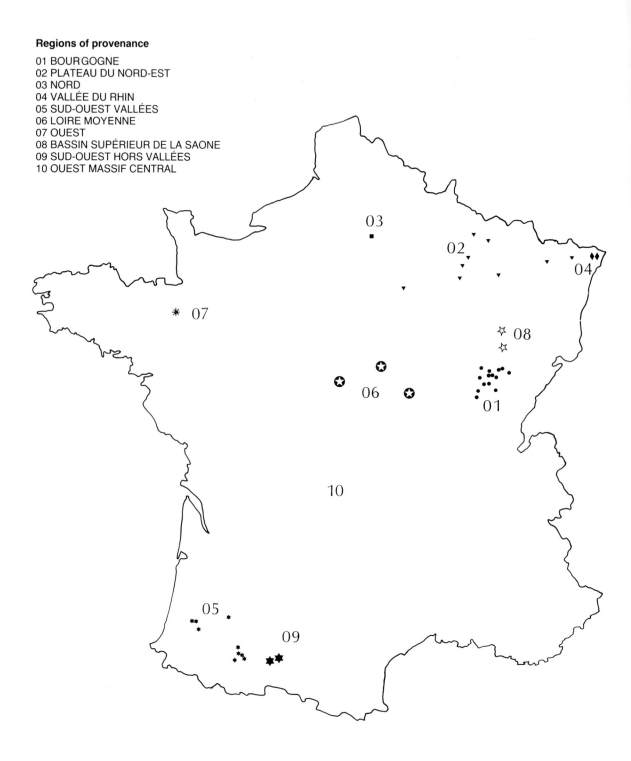

Figure 4.8a Regions of provenance for *Quercus robur* in France.

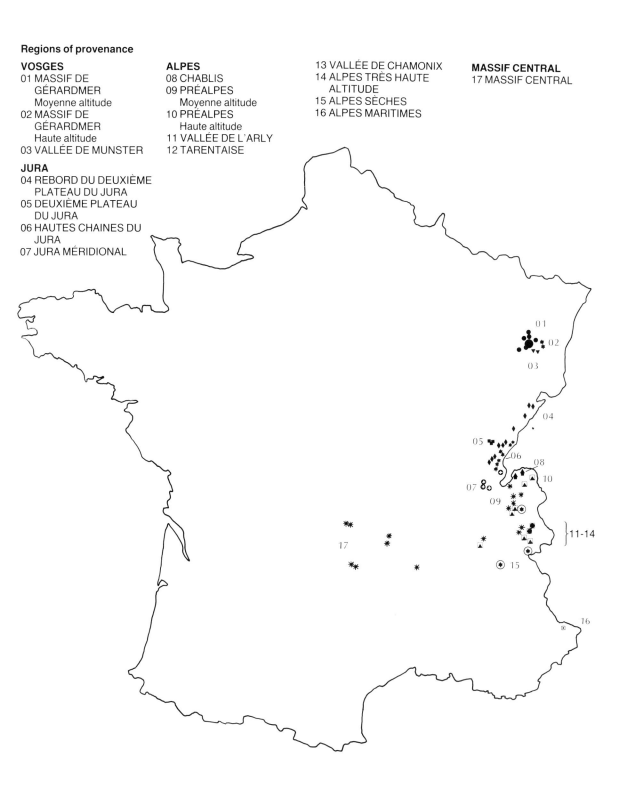

Regions of provenance

VOSGES
01 MASSIF DE
   GÉRARDMER
   Moyenne altitude
02 MASSIF DE
   GÉRARDMER
   Haute altitude
03 VALLÉE DE MUNSTER

JURA
04 REBORD DU DEUXIÈME
   PLATEAU DU JURA
05 DEUXIÈME PLATEAU
   DU JURA
06 HAUTES CHAINES DU
   JURA
07 JURA MÉRIDIONAL

ALPES
08 CHABLIS
09 PRÉALPES
   Moyenne altitude
10 PRÉALPES
   Haute altitude
11 VALLÉE DE L'ARLY
12 TARENTAISE

13 VALLÉE DE CHAMONIX
14 ALPES TRÈS HAUTE
   ALTITUDE
15 ALPES SÈCHES
16 ALPES MARITIMES

MASSIF CENTRAL
17 MASSIF CENTRAL

Figure 4.8b Regions of provenance for *Picea abies* in France.

The second two numbers (14) represent the code for the region of provenance in USA, in this case seed zone 403 Darrington. The next set of numbers represent the year of importation of the seed (not the year of seed ripening as in UK) and the sequential number of the importation in that year. The letter C again represents the species, in this case *Pseudotsuga menziesii*. Finally the name of the seed zone can be added. For seed, a certificate is only issued by the inspector if a cutting test shows the seed to be above a certain standard.

Examples of British identity numbers would be 86(493)I/1, 86(493)II/1, etc.

Denmark (UDC Number 489)

Denmark has identified only one region of provenance for the whole country for all species. Seed stands are selected according to the usual EEC criteria, but in addition to the 13 obligatory EEC species, *Abies grandis*, *Abies procera*, *Acer pseudoplatanus*, *Fraxinus excelsior* and *Tilia cordata* have also been included on a voluntary basis in the certification scheme. However, the OECD scheme (see Part III of this chapter) is also operated for a further seven species of conifers and broadleaves for source identification purposes only (see Anon., 1982).

Each stand that is suitable for registration is given a number, e.g. 426, 427 *et seq.* which is also given to the seed lot. These numbers are allocated sequentially independent of the species and they are used to identify the seed throughout the course of processing, storage, marketing and as plants in the nursery. They are preceded by F (seed stand) or Fp (seed orchard).

In addition to the seed stand number, every seed lot collected and certified in a year is given a further serial number followed by the year, for example 1/81, 2/81 *et seq.* irrespective of species. These are used particularly to identify the year of seed collection and also follow the seed throughout its history and into plants. They are also applied to imported seed lots.

Unlike other countries Denmark also operates a certification scheme for production of plants for environmental use. Stands of trees and ornamental shrubs are registered in just

the same way as for forest use but the certification is carried out by a different agency using different selection criteria.

Since Denmark recognises only one region of provenance (not numbered) the British identity number would continue to use only the UDC number, e.g. 85(489).

France (UDC Number 44)

The French forest authority has decided that the forest area in France cannot be classified into one system of zones of provenance which would be appropriate for all species. Instead, for each species, stands regarded as genetically distinct have been selected in various parts of France and constitute individual zones of provenance for that species. Such zones are numbered consecutively using two digits. Normally, where several selected stands have been identified in one area of France, these have been grouped together under one name and number. These groups are different for each species and no boundaries for these groupings have been delimited (see Figure 4.8a and b). Individual maps of France identifying the exact location of each selected stand have been published by the forest authority for all the EEC species (CEMAGREF, formerly CTGREF, Groupement Technique Forestier, Domaine des Barres, 45290 Nogent-sur-Vernisson, France).

Examples of British identity numbers for an importation from France would be 86(44)01 or 81(44)08.

Germany (UDC Number 430)

The German Federal Authorities have added five species to their scheme over and above those listed in the EEC Directive (see Table 4.3). These are *Abies grandis*, *Alnus glutinosa*, *Fraxinus excelsior*, *Tilia cordata* and *Acer pseudoplatanus*. For all species the Federal German forest resource has been divided up into different zones of provenance according to ecological criteria. Not all species have the same zones of provenance, see Figure 4.9a and b. In theory all seed from the same region of provenance coming from selected stands should perform equally well. In practice many registered stands are in fact derived from man

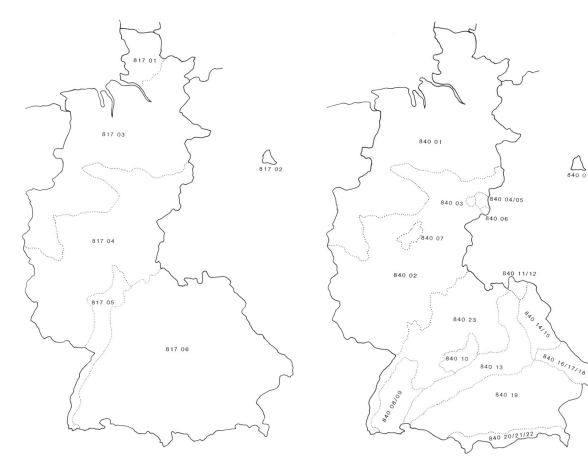

Figure 4.9a Regions of provenance for *Quercus robur* in the former German Federal Republic.

Figure 4.9b Regions of provenance for *Picea abies robur* in the former German Federal Republic.

made plantations and in a large proportion the true origin is not known. For this reason the stands are described as autochthonous (indigenous) or non-autochthonous. Where particular stands are known to be naturally derived, they are much sought after. These are called 'sonder herkunft'. In some zones the selected stands are more homogenous than others, e.g. *Picea abies* 84023 are said to be more homogenous than those in 84019. Each zone of provenance for a different species is allocated a different 5-figure number (see Table 4.3). The first three figures describe the species, the last two the zone. All seed collected from any stand within a zone and the plants grown therefrom are designated by this number.

Since unification, zones of provenance are being extended into the area of the former Democratic Republic, but detailed maps have not yet been published (1991).

An example of a British identity number for *Picea abies* would be 86(430)84001 or 83(430)84019 and for *Quercus robur* 86(430)81703, etc.

Greece (UDC Number 495)

Since its entry into the EEC, Greece has been operating under the transitional arrangements and has been preparing its own set of Regulations. Five regions of provenance based upon biological dry days have been identified. GR – 1/100–150 is Region 1 with 100 to 150 biological dry days; GR – 2/75–100 is Region 2, GR – 3/40–75 is Region 3, GR – 4/1–40 is Region 4 and GR – 5/0 is Region 5. The figures after the / in each case represent the number of biological dry days. No maps of the regions of provenance have yet been published as at March 1991).

Table 4.3 List of species and provenance zones for the former Federal German Republic

Species	Provenance zones*
Acer pseudoplatanus	80101 to 80109
Alnus glutinosa	80201 to 80209
Fagus sylvatica	81001 to 81018
Fraxinus excelsior	81101 to 81105
Quercus rubra	81601 to 81603
Quercus robur	81701 to 81706
Quercus petraea	81801 to 81813
Tilia cordata	82301 to 82307
Abies alba	82701 to 82711
Abies grandis	83001 to 83002
Larix decidua	83701 to 83706
Larix leptolepis	83901 to 83903
Picea abies	84001 to 84023
Picea sitchensis	84401 to 84403
Pinus nigra var. austriaca	84701 to 84703
P. nigra var. calabrica	84801 to 84803
P. nigra var. corsicana	84901 to 84903
Pinus sylvestris	85101 to 85123
Pinus strobus	85201 to 85203
Pseudotsuga menziesii	85301 to 85303

*The location of these zones differs for each species.

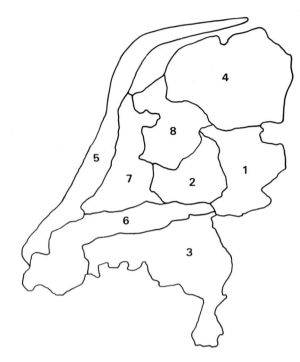

Figure 4.10 Regions of provenance for Holland.

Holland (UDC Number 492)

One set of Regulations has been introduced in Holland to control the certification of forest seed under both the EEC and OECD schemes. The additional species covered in the latter scheme are *Acer pseudoplatanus*, *Alnus glutinosa*, *Alnus incana*, *Betula pubescens*, *Betula pendula*, *Fraxinus excelsior*, *Prunus avium*, *Ulmus glabra*, *Abies grandis*, *Chamaecyparis lawsoniana*, *Pinus contorta*, *Tsuga heterophylla* and *Thuja plicata*.

Holland is divided up into eight main regions based on political and ecological boundaries. These regions are given numbers 1 to 8 and are also named, e.g. Region 8 Zuid Nederland (South Netherlands) (see Figure 4.10). These main regions are further divided into from 0 to 6 subregions which are described by a second number and a name. Thus 3.2 is West Brabant in South Netherlands.

Each sub-region has been further divided into a number of smaller units designated from 01 to 21. These units consist of state forest management areas and private estates and are known collectively by the name of the most important of these in forestry terms. Thus 3.2.04 is the Zundert unit in West Brabant in South Netherlands. These unit names are also the names of the individual seed provenances. Two levels of selection are recognised in Holland, A Selection and 'Ordinary' Selection. A Selection is of a higher quality than 'Ordinary' Selection. Within A Selection provenances, individual stands have been selected and these have been further numbered 01, 02, 03, etc. When material derived from A Selection provenances is described the complete number must be used. Thus NL A.3.2.04.01 is *Quercus robur* A Selection Stand 01 in the Zundert provenance of West Brabant in South Netherlands. However, material from ordinary provenances is now only described by the number of the region provenance, e.g. NL2.

Examples for British identity numbers would be 86(492)1.2 or 86(492)3 for seed from ordinary selections.

Ireland (UDC Number 417)

Only one zone of provenance has been delimited in Ireland for all species. Stands above the national average for yield have been selected and are identified by the National Catalogue Number. This is first a letter designating the species, followed by a number in sequence, e.g. B6 = *Picea abies* for a stand in Dundrum Forest, B7 = *Picea abies* for a stand in Belterbet Forest. Selection of stands is carried out by the Research Branch of the Forest Service, Coillte Teoranta, but the headquarters of Coillte Teoranta is responsible for issuing the certificates. The OECD Scheme is operated for species other than those on the EEC list.

An example of a British identity number would be the same as an old number since only one region is recognised, thus 85(417).

Italy (UDC Number 450)

Forest tree seed certification in Italy is controlled by a special law No.269 dated 22 May 1973 entitled 'Regulations for production and trade of seed and plants for reafforestation'. A comprehensive description of the scheme was published in 1975 entitled *Il materiale forestale di propagazione in Italia* — Collana Verde no.34, 1975 — Ministero Agricoltura e Foreste — Direzione Generale per l'Economia Montana e le Foreste.

The certifying authority is the Italian Forest Service and the selection of seed stands is carried out by Instituto Sperimentale per la Selvicoltura Arezzo with the co-operation of regional forest services. Selection criteria are the same as in the EEC and OECD directives. As in France, the selection unit is the individual stand, which is treated as a separate unit corresponding to a region of provenance. Selected stands are given a progressive identity number in a national list of selected seed stands (Morandini, 1973) irrespective of the species (see Figure 4.11). This number is kept throughout the different operations (collection, handling and storage) and is used when marketed.

Each seed dealer or nurseryman has to keep an official register recording all transactions for

Figure 4.11 Regions of provenance for Italy.

each species and each provenance. Certification is carried out by special representatives of the State Forest Service. In addition to the species listed in the EEC Directives the following are also included in the Italian Certification Scheme: *Abies cephalonica, Cupressus sempervirens, Pinus cembra, Pinus halepensis, Pinus mugo* var. *uncinata, Pinus heldriechii* var. *leucodermis, Pinus pinaster, Pinus pinea, Pinus insignis, Alnus cordifolia, Eucalyptus* spp., *Quercus cerris* and *Quercus suber*.

Examples of British identity numbers for Italy thus take the form of 85(450)20, 86(450)38, etc.

Luxembourg (UDC Number 4359)

No details of the regions of provenance and regulations are yet available.

Portugal (UDC Number 469)

Since joining the EEC, Portugal has been operating under the transitional arrangements. Regions of provenance have been identified and regulations prepared but as at March 1991 no details have been published.

Spain (UDC Number 460)

Since joining the EEC, the necessary regulations to comply with the European Directives have been prepared. Seventeen regions of provenance have been identified. At the same time a scheme complying with the OECD scheme has been drawn up and will soon be in operation. No details of regions of provenance or of the regulations have yet (as at March 1991) been published.

Certification and identification schemes of countries outside the European Economic Community

Argentina (UDC Number 82)

No zones of provenance have been designated nor certification system established so the modified UDC provenance numbers e.g. Neuquen 8281, Tierra del Fuego 8285 would be used. An example of a British identity number would be 85(8281).

Austria (UDC Number 436)

Since 1960 a domestic scheme of registration of forest stands and seed has been in operation. In 1970 the OECD scheme (see p.47) was provisionally introduced and in 1980 it was made compulsory. The Certifying agency is the Forstliche Bundesversuchsanstalt but OECD certificates are only issued for export purposes. (See Anon., 1975).

The country is divided up into major and further minor regions of provenance which are the same for each species and which are normally divided again into altitudinal zones (see Figure 4.12). A series of stands of OECD 'Selected' standard has been identified and each has been given a reference number, for example Fi 37 (VI/1/6−9). Fi = Fichte − *Picea abies;* 37 = Number of the stand from the National Register; VI = Major Region; 1 = Minor Region of provenances; 6−9 = Altitude zone, 600−900 m. These stands were granted equivalence with the EEC in 1983. Seed from stands within the same region of provenance can be bulked together. An example of a British identity number would read 85(436)III/1.

Canada, British Columbia (UDC Number 711)

The only part of Canada of real interest to the UK is British Columbia (BC).

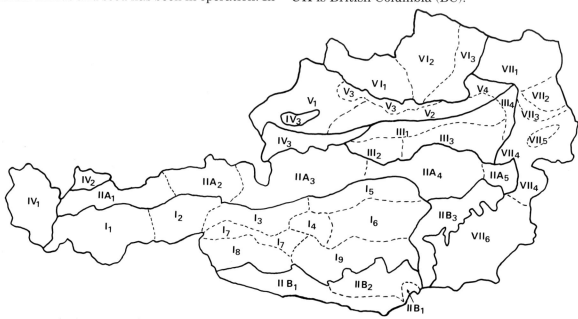

Figure 4.12 Regions of provenance for Austria.

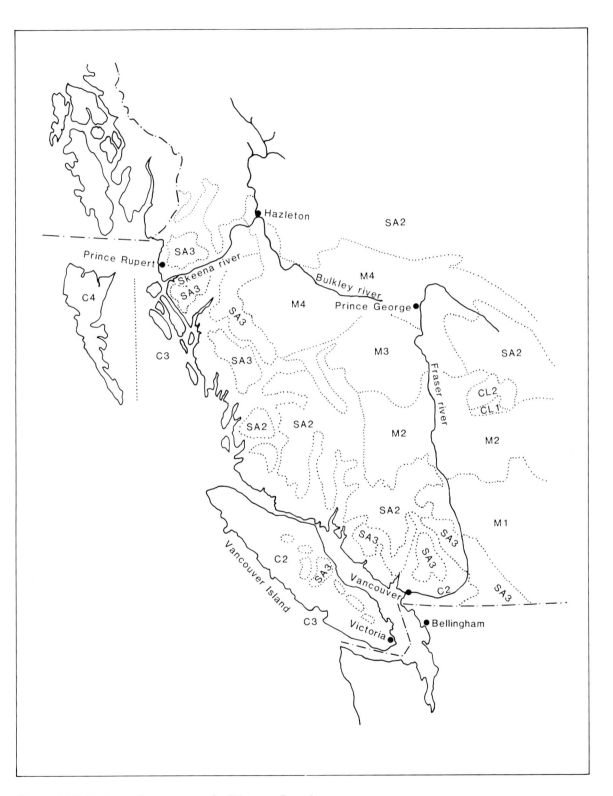

Figure 4.13 Regions of provenance for Western Canada.

The Federal Canadian Forest Service operates the OECD scheme and has divided BC into five forest regions: Boreal (B); Montane (M); Coastal (C); Sub Alpine (SA) and Columbia (CL). These are further divided into a total of 22 sections (see Figure 4.13). These figures are given on the OECD certificates. No attempt has yet been made by the Canadians to select stands according to EEC standards for special collections for export. (See Piesch and Stevenson, 1976.)

Examples of British identity numbers would thus read 86(711)B9, 86(711)C3.

Chile (UDC Number 83)

No zones of provenance have been designated. The modified UDC provenance numbers, e.g. Coquimbo (8317), Nuble (8327) would be used.

Examples of British identity numbers would therefore be 86(8327), 85(8331).

Czechoslovakia (UDC Number 437)

In Czechoslovakia all forest stands more than 50 years in age are classified by their economic value and divided into four categories, IIA, IIB, IIC, IID. The first two categories include those forest stands whose value exceeds the average and are therefore approved for seed production. Stands of category IIA are stands of extraordinary value, mostly indigenous and cannot be felled. Category IIB includes forest stands which are normally felled and regenerated but nevertheless are primarily for seed harvest. Seed orchards are designated in Category I and comprise currently some 5 per cent of the seed production.

Czechoslovakia is divided on a geographical basis into ten silvicultural regions which are used for the harvesting and utilisation of all forest species, and which serve as regions of provenance. In addition the regions are further divided according to the length of the growing season. The basis used is the so-called climatic degree which identifies four zones. These are:

1. high mountain area with a growing season shorter than 100 days;
2. mountain areas with growing season 100–129 days;
3. mountain areas with growing season 130–165 days;
4. lowland and hills with growing season longer than 165 days.

Certification of forest stands for seed is carried out by a National Committee based upon suggestions from the local forest managers. Suggestions for inclusion as Category IIA are referred to the Forest Research Institute where the whole seed certification scheme is co-ordinated. Selected seed stands are given a number, an abbreviation for the tree species, the category of the stand, the silvicultural region and the administrative district, for example, IIB 720 V SU. Each year each seed lot is given a simple number in sequence, e.g. 431/81;49/83. These numbers appear on the Certificates issued by the Forest Authority. However, they do not determine regions of provenance. These have now been identified for Slovakia for *Abies alba, Larix decidua, Picea abies* and *Pinus sylvestris*. None has so far been published for other areas. The region is not included on the Certificate of Provenance. They can only be worked out from the place names or forest districts found on the Certificates.

Typical British identity numbers would be 86(437)1 or 86(437)9.

Hungary (UDC Number 439)

Hungary has been operating the OECD scheme since December 1989 and has delimited nine regions of provenance (10, 20, ... 90). These are based upon six geomorphological or climatic regions, with the sixth region (the Great Plain) being divided up into four regions, 60, 70, 80 and 90 according to the soil type. There are therefore discontinuities within all but two regions (20 and 30). Each region is further subdivided into up to 9 sub-regions (11, 12, ... 19; but 71, 72 and 73 only) corresponding to forest regions of similar forest cover (see Figure 4.14).

The species covered by the OECD Scheme are *Quercus robur. Q. petraea, Fagus sylvatica, Carpinus betulus, Fraxinus excelsior, Robinia pseudoacacia, Populus* spp., *Salix* spp., *Pinus sylvestris, Larix decidua and Picea abies.* All

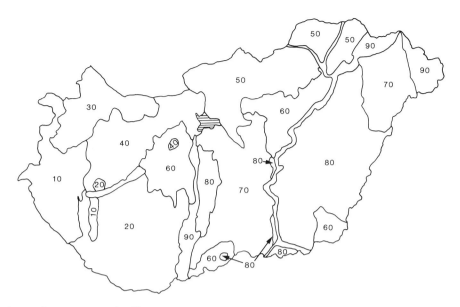

Figure 4.14 Regions of provenance for Hungary.

OECD categories are covered and tested seed orchard material of *Pinus sylvestris* and tested clones of *Populus* spp. and *Salix* spp. are already commercially available.

Japan (UDC Number 520)

Although a certification scheme following closely the OECD pattern has been in operation since 1970, the Japanese authorities have not allowed seed from selected sources to be exported until their own needs have been satisfied. In practice this has meant that little or no certified seed has been available for export and the only source certificates available are those from the local Chamber of Commerce, e.g. Nagano or Hokkaido.

Poland (UDC Number 438)

Since 1952 Poland has been divided up into eight botanical and forest regions based on species distribution, climatic conditions and soil characteristics. These are designated by Roman numerals I to VIII. For any one species from one to seven smaller areas have been identified throughout the distribution where particularly good quality stands of that particular species have been located. These are given arabic numbers.

These regions differ for each species mentioned and may be likened to regions of provenance. The eight main regions are shown in Figure 4.15. Of these, the only ones of interest to British forestry (see Chapter 3) are Region VII, Sudety, and Region VIII, Karpaty. In the former, the Sudeten origins of *Larix decidua* are indigenous, whereas in Region VIII stands of *Picea abies* around the Istebna forest, which have done well in British trials, are to be found.

Figure 4.15 Regions of provenance for Poland.

The stands of particularly high quality in these regions are equivalent to the selected category in EEC countries. In fact, in order to qualify for possible implementation of the OECD scheme, the standards of selection of these stands have been brought into line with those of the OECD scheme. At time of writing Poland has formally applied to be allowed to operate the OECD scheme.

Further stands of a slightly lower quality within the natural distribution of each species, as well as good quality stands of non-indigenous origins, have also been selected and these have been given a lower case letter – a to d for *Picea abies,* and a to e for *Pinus sylvestris,* a to c for *Quercus robur* and *Fagus sylvatica,* d for introduced stands of *Larix leptolepis,* and a to c for *Larix decidua* var. *sudetica.*

The method of identifying seed from these stands utilises the numbers and/or letters plus the names of the capital of the forestry administrative region and of the actual primary forest from whence the seed came. Thus seed from selected stands of Istebna *Picea abies* would be designated *Picea abies* VIII/5 Karpaty Krakow/Katawice, Istebna beslivo.

Examples of British identity numbers would be 86(438)I or 86(438)VI.

Romania (UDC Number 498)

Since 1981 OECD certificates have been available for all the important forestry species exported from Romania. The implementation of the scheme followed closely the publication of a new system of seed zones (Enescu and Donita, 1976). This divided the country into 15 different seed *Regions* based on geographical and ecological criteria (see Figure 4.16). These are classified A to O. For example A = northern part of the Eastern Carpathians, B = Southern part. These are further sub-divided into up to 4 *sub-regions* based upon geographical limits such as watersheds. These are given the numbers 1–4, for example A_1 = western watershed, A_2 = eastern watershed. Further sub-divisions are made into up to nine *sectors* based upon vegetation and are ecologically very homogenous, for example, $A_1 1$ *Picea abies*, $A_1 2$ mixed forest of beech and conifers, $A_1 5$ sessile oak forest. In the majority of cases one sector is used as a seed zone but each sector may be

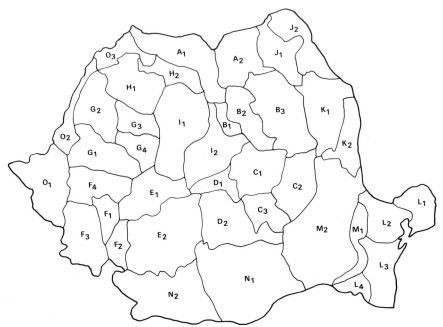

Figure 4.16 Regions of provenance for Romania.

divided up into three elevation bands of not more than 400 metres if the range of the natural forest vegetation of the zone exceeds this. Where no elevation levels exist this is designated by zero, e.g. $A_1 50$, otherwise the upper elevation band is designated 1, the intermediate 2 and the lowest 3. For example $A_2 11$ is 1300–1600, $A_2 12$ is 900–1300 and $A_2 13$ is 700–900. Within each of these seed zones at least one stand of above average quality has been selected for seed collection in order to give enough seed for use within each zone. Seed from each zone and elevation, where applicable, is used as a priority over seed from another zone. No other seed may be used in Romania. All seed from Romania will be designated thus $R/81/B_2 11$ (Romania/Year of collection/Region, sub-region, sector and elevation).

In addition to the above information OECD Certificates of Origin, issued by the certifying authority of the Forestry Department, will include a site index. This gives an indication of the yield potential of the site and can range from I to VI. Seed stands have only been selected from site indices I and II.

Examples of British identity numbers would read $86(498)A_1$ or $86(498)A_2$. (Subscripted number will appear full sized in computer listings.)

Turkey (UDC Number 560)

For some time a system of forest classification has been in operation in Turkey. This has included a source-identified certification scheme widely used for *Abies nordmanniana*. An OECD scheme has been prepared and Turkey has intimated its intention of applying to operate the OECD scheme in the near future (information as at March 1991).

USA (Most important UDC Numbers: Alaska 798, Washington 797, Oregon 795 and California 794).

In 1966 the merchants of the Pacific North West, recognising the importance of seed origin to their European customers and being aware of the poor reputations some of their firms had built up, formed themselves into the Northwest Tree Seed Certifiers Association (NWTSCA) and began issuing certificates. These were of three kinds, Audit Class, Source Identified A and Source Identified B. These gave increasing levels of authenticity to the seeds. At the same time they divided the States of Washington and Oregon into a large number of seed zones based on climatological criteria. The complete zones are published in map form by the Northwest Tree Seed Council (see Figure 4.17). Each zone was given a 3-figure number based on the broad ecological boundaries of main watersheds, subsidiary valleys, aspects and forest distributions, for example 030, 403, etc. Collections were made in elevation bands of 500 feet which was designated by 05, 10, 15, etc. 05 indicated the elevation band 0–500 feet *et seq*. Since the zones were first published a few minor changes have been made with some of the large zones being further split. In 1974, in response to approaches from the EEC, the NWTSCA modified their procedures slightly to allow their own scheme to operate within the OECD scheme. The current schemes for Washington and Oregon are published separately by the Certifying Authority of each State. A summary of both schemes was written by Hoekstra (1976). Other states in the USA are designated only by their UDC number.

The three certification classes of seed so far available only pertain to source identification. As yet only very minor attempts have been made to select stands for physical quality. The various classes are as follows:

Certified seed – seed from trees of proven genetic identity, as described in the Certificate of Genetic Identity. Such seed is produced and processed in a manner assuring genetic identity common with the tested material and will be labelled with a blue label stating 'Certified seed.'

Selected seed – seed from trees having promise of, but not proof of genetic identity, as described in the Certificate of Genetic Identity. Such seed shall be labelled with a green label stating 'Selected seed'.

Source identified seed – seed from within a seed zone or portion thereof and from within a 500-foot elevation increment. The tree from which the seed is collected is assumed to be indigenous. Two subclasses are recognised:

Subclass A: personally supervised collections; Subclass B: procedurally supervised collections.

Both subclasses are labelled with a yellow label stating 'Source-identified seed'; and the subclass.

SUBCLASS A

– source identified seed means that the applicant and certifying agency personally know beyond a reasonable doubt the seed zone or portion thereof and 500-foot elevation increment within which cones and/or seed were collected. The Certifying agency knows the location from the applicant's prior written plan of his cone collecting activities.

SUBCLASS B

– source identified seed means seed identified as collected from within a seed zone or portion thereof and from within a 500-foot elevation increment. The Certifying agency does not necessarily carry out a field inspection.

– audit certificate seed means that the applicant's record of procurement, processing, storage, and distribution state that the seed was collected from within stated seed zones or described portions thereof and from within 500-foot elevation increments. Containers of this seed carry a serially numbered brown and white label stating 'Audit certificate seed'. All records of the applicant for this class of seed are subject to audit.

All certificates issued use the number of the seed zone and elevation to identify the seed source, thus 403-05, 632-15. These zones are used widely in most other EEC countries and will be found as the identification on certificates issued by other EEC member states. Because of their wide application the Universal Decimal Classification used by the Forestry Commission until 1984 is also given (see Figure 4.17), so that it will be possible for nurserymen to relate to the two systems at a glance.

Identity numbers for Washington would have the form 86(797)403, 86(797)030, etc., and for Oregon 81(795)051, 83(795)461, etc.

This OECD system of certification is not applied in Alaska. This state is divided up into 27 provincial seed zones along the coast which are numbered 1 to 27. Examples of British identity numbers would be 85(798)1 or 86(798)26.

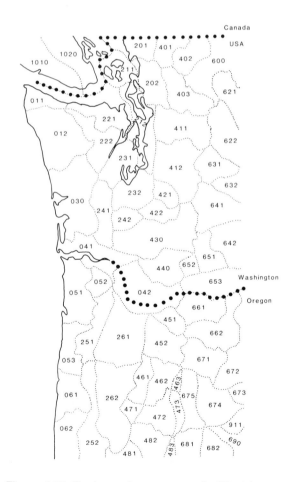

Figure 4.17 Regions of provenance for Washington and Oregon.

The OECD scheme for the control of forest reproductive material moving in international trade

Introduction

During the 1960s the wish that planting stocks for forest use should be from authenticated sources of seed and of the best possible genetic quality, led the Organisation for Economic Co-operation and Development to draw up a Seed Certification Scheme for forestry similar to schemes established previously for agricultural cereal and crop seeds moving in international trade. The scheme was published in 1974. Since the scheme was likely to be used in parts of the world where selected stands are not yet available, a category, 'Source identified', was included. In addition, because several countries possess untested seed orchards which do not meet the standards set by the EEC, for example, an additional category – 'untested seed orchards' – was also included.

The most recent version of the scheme lays down guidelines as to the degree of supervision necessary at all stages from collection through processing, storage and distribution of seeds that must be adopted by a Certifying Authority (OECD, 1976). A member country of the OECD wishing to apply the scheme has to submit copies of the rules and the labels and have these approved before they are entitled to start operating it. The procedures are similar to the standards laid down in the EEC Directive. The scheme can be applied to any species. Consequently, the rules of the OECD scheme have been adopted by several EEC countries to supplement the legally enforceable scheme based on the EEC Directive.

Certain non-member countries of the OECD with a large external market for seed have recognised the advantages of applying the OECD scheme and to date, Romania, Hungary and Poland have formally applied to be allowed to operate the scheme. Representatives of the OECD visited Romania and Hungary and recommended that the standards of control of collections and processing were sufficiently high to warrant their adoption of the OECD scheme. As a result in 1981 Romania and in 1989 Hungary were given permission to begin selling seed certified under the OECD rules. Currently the Polish application is still under consideration.

The OECD scheme in Britain

Because the Forest Reproductive Materials Regulations do not cover all forest species grown in the United Kingdom, and because seed from some of these not covered is sold abroad, it was decided to implement the OECD scheme in Britain so that plants of all species can be officially certified, provided the proper procedures have been followed.

A leaflet *Scheme for the control of forest reproductive material moving in international trade* was published in 1982 and took immediate effect. It describes the procedures to be followed if plants and seeds ·are to be certified and sold abroad under the OECD label. The rules parallel very closely the procedures contained in the Forest Reproductive Material Regulations 1977. The booklet outlining those Regulations, published by the Forestry Commission, also includes details of the OECD scheme.

REFERENCES and FURTHER READING

ANON. (1965). *Seed identification numbers.* Forestry Commission Research Branch Paper 29. (370 pp.). Forestry Commission, London.

ANON. (1975). *The Austrian Forestry Act No. 440 of 3 July 1975, Chapter XI – Forest seeds and plants.*

ANON. (1982). *Approved seed stands in Danish forests.* Klampenborg.

ENESCU, V. and DANITA, N. (1976). Zonele de recoltare a semintelor forestiere in RD Romania. *ICAS,* Ed. Ceres, 1-31 & annex.

FORESTRY COMMISSION (1989). *Native pinewoods – grants and guidelines.* Forestry Commission, Edinburgh.

HOEKSTRA, P.E. (1976). *Certification of source identified tree seed under the OECD scheme in Oregon and Washington, USA.* USDA Forest Service, 1-14.

MORANDINI, R. (1973). Libro nationale boschi da Seme. *Annaii dell'istituto sperimentale per la Solvicoltura,* Arezzo. **IV.** Italy.

OECD (1976). *The OECD scheme for the control of forest reproductive material moving in international trade.* Organization for Economic Co-operation and Development.

PIESCH, R.F. and STEVENSON, R.E. (1976). *Certification of source-identified Canadian tree seed under the OECD scheme.* Forestry Technical Report 19. (18pp.) Canadian Forestry Service, Ottawa, Canada.

Chapter 5
Seed Supply and Aspects of the Law

by **A. G. Gordon** *and* **J. R. Aldhous**

Introduction

This chapter outlines the three most immediate legal constraints affecting collection and trading in seed:

the Forest Reproductive Material Regulations, 1977; the Plant Health (Great Britain) Order 1987, as amended; the Plant Health (Forestry) (Great Britain) Order 1989.

Brief advice is also given on securing civil permission prior to seed collecting.

There is, in addition, a short section on seed testing. Other legislation that might in certain circumstances have a bearing on seed trading (e.g. health and safety; control of pesticides; trade descriptions and other consumer protection legislation, etc.) are not discussed.

Forest Reproductive Material Regulations

The control of the quality of material used for forest plantings in Great Britain is exercised through the Forest Reproductive Material Regulations 1977 (Statutory Instrument 1977 No. 891). These Regulations which are based upon Directives 66/404, 71/161 and 75/445 of the European Economic Community, apply to the marketing of reproductive material (seed, cuttings and plants) for forestry purposes and cover not only the genetic quality of the parent material from which the reproductive material is derived, but also the physical quality of the material to be marketed. Any contravention of the Forest Reproductive Material Regulations may be an offence under Part II of the Plant Varieties and Seeds Act 1964.

The details of the Regulations that apply to the physical quality of seed are to be found in Chapter 11 on Seed Testing in this Bulletin. This chapter outlines the scope of the Regulations only insofar as they apply to the registration and marketing of seed; it is not legally authoritative and reference should be made to the Regulations on matters of detail and legal definition.

The Regulations apply only to the species listed in Table 5.1. These are commonly referred to as the EEC species and include some which, although of little commercial forestry importance in Britain, were in the European Directives, being of commercial importance in the original six European Community countries. Some member countries have taken advantage of the provision in the Directives to supplement the list so as to control additional species within their domestic market.

The Directives are currently under revision, partly to respond to the needs of the enlarged Community, now including Greece, Portugal and Spain, and partly to take account of advances in tree breeding.

The Regulations do not apply to seed or cones for export to non-EEC countries or to seed lots authorised by the Forestry Commission in writing to be used in tests or for scientific purposes. Nor currently (1991) do they apply to the use of plants of the EEC species for purposes other than forestry (e.g. landscaping).

An explanatory booklet (1987) for *The Forest Reproductive Material Regulations 1977* has been prepared and is available free of charge

Table 5.1 Species to which the Forest Reproductive Material Regulations (1977) apply

Common name	Botanical name	Synonym
Silver fir	*Abies alba* Mill.	*Abies pectinata* DC
Beech	*Fagus sylvatica* L.	
European larch	*Larix decidua* Mill.	
Japanese larch	*Larix leptolepis* (Sieb. & Zucc.) Gord.	
Norway spruce	*Picea abies* Karst.	*Picea excelsa* Link.
Sitka spruce	*Picea sitchensis* Trautv. et Mey	*Picea menziesii* Carr.
Austrian and Corsican pine	*Pinus nigra* Arn.	*Pinus laricio* Poir.
Scots pine	*Pinus sylvestris* L.	
Weymouth pine	*Pinus strobus* L.	
Douglas fir	*Pseudotsuga taxifolia* (Poir.) Britt.	*Pseudotsuga douglasii* Carr. *Pseudotsuga menziesii** (Mirb.) Franco
Red oak	*Quercus borealis* Michx.	*Quercus rubra** Du Roi
Pedunculate oak	*Quercus pedunculata* Ehrh.	*Quercus robur** L.
Sessile oak	*Quercus sessiliflora* Sal.	*Quercus petraea** Liebl.

*These synonyms are commonly used in British forestry and have therefore been used throughout this Bulletin.

from the Grants and Licences Division, Forestry Commission Headquarters, Edinburgh. It attempts to explain the legal requirements of the Regulations and gives more specific information than will be found in this chapter.

Registration

To meet the requirements of the Directives each EEC member country must compile and maintain its own National Register showing approved sources of basic material within their country. (*Basic material* is the EEC term used to define seed stands and seed orchards and poplar varieties propagated vegetatively.) The British Register is maintained by the Forestry Authority and opies are available for inspection at Forestry Commission Headquarters in Edinburgh, the Official Testing Station for Forest Reproductive Material, Alice Holt Lodge, Farnham, Surrey and at each Forestry Authority National Office.

Registration procedures (Regulations 4, 5, 6 and 7) require that an owner seeking approval and registration for a seed stand must apply in writing to the appropriate Chief Conservator who, on receipt of a fee (which is higher for any applications made near to harvest), will arrange for the stand to be inspected. If approved, it is entered in the Register under one of two categories, namely:

selected (which are stands found to be suitable for reproductive purposes having no characteristics undesirable for forestry purposes. These stands are commonly referred to as Registered Stands and the seeds derived therefrom as Registered seed);

tested (as selected but additionally showing improved value for use for forestry purposes as determined by tests defined in the Regulations). 'Tested' currently only applies to poplar varieties and to seed from a limited number of seed orchards.

It is very likely that a third category, *untested seed orchards*, will be introduced following revision of the EEC Directive. This will include orchards in the 'NT' category.

A typical entry in the British Register is shown as Appendix 5.1.

Machinery is provided for appeal against a decision to refuse registration or to remove from the Register stands already registered.

Although countries that are not members of the EEC may operate their own systems for recording seed sources and produce what is often called selected seed (see Chapter 4) the term Registered under the Regulations refers only to stands within the EEC which have been entered in a National Register maintained by a member State. All other stands whether within or outwith the EEC are considered to be Unregistered, unless granted equivalance (see Austria, p.40).

Under the Regulations, the Forestry Authority has prepared a map delineating regions used when describing the provenance of registered seed. Details of the regions and identification system used are found in Chapter 4. Maps of the regions of provenance are held at Forestry Commission Headquarters, at the Official Testing Station and at each Forestry Authority National Office.

Collection, marking and labelling

Any person proposing to collect cones or seed of any of the EEC species which (or the derivatives of which) he intends to market within the EEC, is required to inform the appropriate Forestry Authority National Office in writing 28 days in advance, of the place of collection and the proposed starting and completion dates (see Regulation 8). At the discretion of the National Office staff, this period of notice may be shortened. Normally such collections will be made only from registered stands. However, an owner may collect cones or seed from an unregistered stand provided he does not market within the EEC the seeds or any plants raised for forestry purposes form such seed. In some circumstances, for example when there is a shortfall of registered seed of the appropriate origin within the EEC, permission may be given in writing by the Forestry Authority for unregistered seed to be marketed for forestry purposes (see below).

During collection the owner of the stand or his agent must ensure to the satisfaction of the Forestry Authority that the material being collected is from the declared stand and on completion must inform the appropriate Chief Conservator of the quantity of cones or seeds collected; at the end of extraction the quantity of seed extracted from the cones must also be notified. Failure to comply may mean that the appropriate Certificate of Provenance (see below) – without which the seed cannot be legally marketed – will be withheld.

Before the seed can be marketed it must also be tested for its physical quality (see below).

During collection, processing and marketing, seed and cones must be kept in separate lots distinguished from each other by reference to species, category, provenance, origin and year of ripening.

Each lot from a Registered stand must be marked by a green label and each lot from a tested source must have a blue label (see Regulations 9 and 10). Specimens of suitable labels are shown in Appendix 5.2.

If heavy seeding or coning is observed on unregistered stands of high quality and the species is subject to Forest Reproductive Material Regulations, the owner or seed collector can apply to have the stand added to the National Register. It is always desirable that the application is made as early as possible. Late applications may necessitate payment of an increased Registration fee.

Marketing

Regulation 11 requires that seed may only be marketed if it is from a Registered stand. However, it does allow Unregistered seed to be marketed provided it is covered by a Forestry Authority licence. Such licences are issued (subject to specified conditions and to a derogation decision made in Brussels) if the Forestry Authority is satisfied that insufficient registered seed of the appropriate origin is available within the EEC and authority has been received from the European Commission in Brussels to permit 'marketing of basic material meeting less stringent requirments' than are set out in the Directive. The licence also authorises plants grown from seed to be marketed.

The regulation stipulates that no registered seed shall be marketed except under the description 'EEC Standard' and unless its container has been sealed with a non-reusable seal. It also stipulates that no seed may be marketed unless it is covered by a valid Seed Test Certificate (or in the case of seed imported from a member state or Northern Ireland, by documentary evidence) that it does not contain more than the permitted quantity (detailed in Schedule 7, part III of the Regulations) of seed of other forest species. However, the Forestry Authority has the powers to authorise the marketing of seeds which do not meet the EEC Standard, if the situation warrants it. In Britain a Seed Test Certificate can only be issued by the Official Forest Seed Testing

Station at Alice Holt Lodge to which samples drawn according to the proper procedure should be sent (see below).

Anyone who sells any seed covered by the FRM Regulations is also required by Regulation 12 to forward to the purchaser within 14 days a Supplier's Certificate. This document must contain details of the physical quality of the seed as reported in the Test Certificate as well as details of the species, type, quantity, category, provenance and origin. Supplier's Certificates must be printed on green paper for selected material and for tested seed on blue paper. In order to differentiate between Registered and non-Registered material, suppliers in Britain are in the habit of using white Certificates for the latter.

For some species which cannot be stored satisfactorily without significant loss of quality and in some other circumstances, this requirement to provide a Supplier's Certificate within 14 days has proven impractical to implement. Thus, for example for *Quercus* spp. and *Fagus sylvatica* required for autumn sowing, it has been accepted that Supplier's Certificates should be forwarded as soon as possible after delivery. This ruling has also applied for those conifers required immediately after processing.

Small quantities of seeds (defined as the quantity sufficient to produce not more than 1000 usable plants) which are not intended for forestry purposes, and seeds authorised by the Forestry Authority to be used in selection work, are exempt from certain regulations relating to the need for control of collection and issue of Supplier's Certificates.

Importation

One of the main objectives of the EEC Directives on the Forest Reproductive Material is to encourage the marketing of tree seed between member states and, provided registered seed from ecologically comparable sources is available within the EEC, it should be used in preference to seed imported from other countries. However, several practical difficulties in the supply and demand of registered seed within the EEC have, in the past, necessitated importation from countries outside the EEC.

The Directives make provision for marketing unregistered imported seed provided authority has previously been given by the European Commission. This authorisation, informally known as derogation, takes account of each member state's contingency requests to market seed from unregistered stands both within and outwith the EEC should there be a shortfall of registered seed in the following season. The Forestry Authority is responsible for making the British derogation submission, which it does according to a timetable set by the European Commission after consulting the relevant trade organisations and seed merchants. The decision of the European Commission is usually given within 2–3 months.

Appendix 5.3 shows the documents required when importing seed for marketing. For reasons already mentioned under the marketing section, seed may be despatched before full test results are available. Any importer faced with difficulties in gaining Customs clearance through lack of a test certificate should seek advice from the Grants and Licences Division at Forestry Commission Headquarters.

Import licences are available only from the Forestry Authority. Applications, which should be sent to the Grants and Licences Division, will normally only be granted for species on the EEC list if prior submission has been made for derogation (see above). They may also impose certain conditions. If the full quantity for which Britain sought derogation has not been taken up a licence may be granted even if no prior submission has been made by the would-be importer. It is also possible to submit a late application for derogation if the permitted quantities have been exceeded.

The regulations pertaining to importation (regulations 14 and 15) do not apply to small quantities of seed (sufficient to produce not more than 1000 usable plants) to be used for purposes other than forestry, and to seed or cones authorised by the Forestry Authority to be used in tests (e.g. samples of seed sent for prior testing in Britain) or for scientific purposes or selection work.

Plant health requirements

Plant Health legislation in Great Britain, as in other Member States in the EEC, is based on the requirements of the EEC Plant Health Directive 77/93/EEC, as amended. The enabling legislation in this country is the Plant Health (Great Britain) Order 1987 (Statutory Instrument 1987 No. 1758, as amended by SI 1989 No. 553) and the Plant Health (Forestry) (Great Britain) Order 1989 (Statutory Instrument 1989 No. 1951). This SI, which has more bearing on forestry material, revokes the Import and Export of Trees, Wood and Bark (Health) (Great Britain) Order 1980 and its amending Orders (SI 1980/449 as amended by SI 1983/807, SI 1984/1892 and SI 1986/196).

As a general guide, the definitions of 'plant' and 'tree' in the Plant Health Orders includes fruit and seed. Where controls are in force in respect of any species of plant or tree, unless specifically excluded in the appropriate Schedule to the Order, those controls extend to fruit and seeds. Thus, for example, seeds of *Quercus* and *Castanea* fall within the scope of the regulations and importations of seed of these species must be accompanied by a phytosanitary certificate certifying that the area of origin is known to be free from *Cryptonectria (Endothia) parasitica* and *Cronartium quercum* for *Quercus* spp., and of *C.parasitica* for *Castanea* spp. (It should be noted that this differs from the interpretation previously reported in Bulletin 59.)

Although strictly speaking outwith the scope of this Bulletin it is worth pointing out that the seeds of some species of ornamentals, not generally used for forestry purposes, are also controlled (SI 1987 No. 1758). Imported *Prunus,* for example, requires an appropriately worded phytosanitary certificate because of the risk of introduction of certain harmful organisms which attack agricultural or horticultural crops.

Plant Health requirements in respect of imports and exports of fruit and seed covered by the legislation are complex and to assist those involved in such trade, the Ministry of Agriculture, Fisheries and Food has published its *Guide for Importers – Plant Health Legislation* which is updated from time to time. The Guide, which gives a tabular breakdown of the requirements of both Plant Health Orders, also gives contact names and addresses; importers with specific queries are advised to check with the responsible authority to establish the precise position. This will reduce the risk of delays at Customs points through non-compliance.

Finally, note must be taken of the advent of the Single European Market in 1993. Already, proposals are being put forward within the EEC Commission which indicate how significant the effect will probably be on plant health matters. Undoubtedly further changes in the legislation will follow. For an up-to-date report at any time reference should be made to the Plant Health Branch, Forestry Commission Headquarters or MAFF Plant Health Advisory Unit, Nobel House, 17 Smith Square, London SW1P 3HX (telephone 071-238 6477).

Export of seed

There are no export restrictions imposed by the UK authorities on registered seed within the EEC. Both it and unregistered seed which has derogation approval cannot automatically be marketed in other member countries, and in practice may not be allowed in by the forestry authority of the importing EEC country due to inappropriate origin, etc. Seed of the EEC species sold to member countries must always be accompanied by a Certificate of Provenance and a current Seed Test Certificate. A Supplier's Certificate (which is signed by the exporter) may not be accepted by the forest authority of the importing member country but should be supplied in order to comply with the British Regulations.

53

The effect of the 1993 reduction of trade barriers within the Community should facilitate international trade in seed.

Certified copies of the original Certificate of Provenance and Seed Test Certificate may be acceptable to the importer but if these are refused new certificates can be obtained from Forestry Commission Headquarters and the Official Seed Testing Station.

There are no export restrictions on non-EEC species to some member states but others, including Holland, Belgium, France, the Federal Republic of Germany and Italy have extended their regulations to include other species and Certificates of Provenance and Seed Test Certificates may be required before the seeds of these species will be allowed entry into these countries.

Although Seed Test Certificates are required by the EEC Directives most member states have recognised that this is not always possible to provide in practice and have therefore allowed seed to enter their country without presentation of a Test Certificate.

There are no British or European restrictions on the export of seed of any species to countries outside the EEC. However, Sweden has recently imposed a restriction on importation of *Pinus contorta* from countries other than Canada and many other countries have phytosanitary requirements if not for Certificates of Provenance also. Would-be exporters are advised to check the requirements of the particular country before despatching the seed. Any person needing to obtain a phytosanitary certificate should apply to his local Plant Health and Seeds Inspector, who will sample the seed lot according to correct procedures. Such samples are then sent to the Ministry of Agriculture's Harpenden Laboratory to be examined for important plant diseases. Once the seed lot has been checked and found to be in a healthy condition, the Ministry will issue a phytosanitary certificate.

Records

The Forestry Authority has advised all those known to them to be concerned in the collection, extraction, processing, storing, transporting, raising or marketing of forest reproductive materials, that they must keep suitable records for 7 years to show that the material in their charge has been controlled adequately at all times. Those licensed to market small quantities of seed for non-forestry purposes are required to send to the Forestry Authority an annual return of sales. The Forestry Authority's nursery inspectors have the right to examine these records and to prohibit the further marketing of any material not adequately controlled. A lack of knowledge of the Forest Reproductive Material Regulations will not normally absolve anyone from the need to keep such records.

Other preliminaries prior to seed collection

Anyone considering organising seed crop forecasting, expecting this to lead to seed collection from specific areas of growing or felled trees, should, firstly, ascertain who is able to give permission and must ensure that he has full permission to proceed. Where seed is to be collected from recently felled trees, the organiser must ensure beforehand, that his and his seed collectors' position is clearly established with any contractor or subcontractor who may have legitimate access to the site for felling, conversion, extraction, etc., or to prepare the site for replanting.

Secondly, the seed collection organiser should ensure that he has adequate insurance cover for all aspects of the operation, should any claim be made against him, for example, by a seed collector injured by felled timber which moves while cones are being gathered, or by falling from a ladder while collecting from growing trees.

Landowners should ensure that any insurance they have against third party claims arising from their woodlands, includes claims that might arise from seed collection.

Seed testing

The 1973 FRM Regulations, which revoked

that part of the Seeds Regulations of 1961 (Statutory Instrument 1961 Nos 212 and 274) relating to forest seed, require Seed Test Certificates to be issued for all seed sold in Great Britain of the 13 species covered by the FRM Regulations but set no requirements for other species previously covered. There are no exceptions to the section (13) of the Regulations which refers to seed testing; Seed Test Certificates are required, whatever the ultimate destination. Any seed of the species listed in Table 5.1 that is sold in Great Britain, whether for ornamental or forestry use, must be covered by a current Seed Test Certificate. If any of these seeds have been stored for a significant time in Britain or if they have been harvested in Britain, the Seed Test Certificate must be issued by the Official Testing Station for forest seed at Alice Holt Lodge, Wrecclesham, Farnham, Surrey. On the other hand if the seed is bought direct from a foreign supplier or if it has recently been imported but immediately remarketed by a British supplier, then the exporter's Seed Test Certificate presented to HM Customs is acceptable, if still current. For further explanation on currency of seed testing and the seed testing year, see Chapter 11. However, it should be noted that, when the amount of seed carried forward to a new seed testing year is small and the value is not many times more than the cost of seed testing, it may be possible to obtain a licence from the Forestry Authority to market such seed using the previous year's test results as the basis for the Supplier's Certificate.

No other commercial tree species are currently covered by seed testing requirements and at present there are no intentions of introducing legislation to cover the various species that were covered prior to 1973. However, sales of seed fall within the normal consumer protection legislation and merchants selling falsely identified or dead seed are legally responsible for their actions. Retrospective claims never completely compensate a nurseryman for lost planting stock, so he should always insist that a current Seed Test Certificate is available for any seed he buys. He would also be well advised to establish minimum purity and germination or viability standards, or alternatively minimum numbers of germinable or viable seeds per kilogram before a purchase is confirmed. It is also in a seed merchant's own interest in the event of a claim to be able to prove by means of a current Seed Test Certificate that the seed he has sold is of a certain standard. See also Chapter 11.

Appendix

National register material produced by sexual means

Region of provenance	40	**Genus**	Pinus
Place of provenance (*if non-indigenous*)	—	**Species**	nigra
		Sub Species	—
Origin	Unknown	**Variety**	maritima
Date of planting	1965	**Identity No.**	4003
Stand No. 4003	Area (ha) 6.0		
Owner	Forestry Commission	**Forest or Estate**	Sherwood
Address	Great Eastern House Tenison Road Cambridge CB1 2DU	**Cpt. or Wood**	CPT 190a
		Conservancy	E (ENG)
		County	Nottingham
Agent	—	**National grid ref.**	SK669749
Address	—		
		Date of registration	August 1973
		Entry No.	7/4
		Date of withdrawal	—
		Reason for withdrawal	—

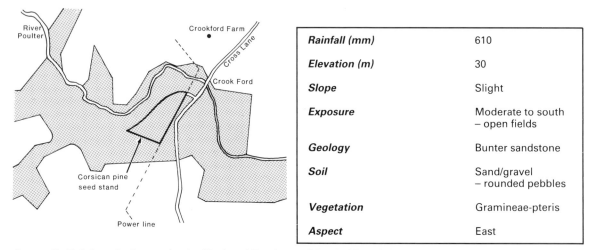

Rainfall (mm)	610
Elevation (m)	30
Slope	Slight
Exposure	Moderate to south – open fields
Geology	Bunter sandstone
Soil	Sand/gravel – rounded pebbles
Vegetation	Gramineae-pteris
Aspect	East

Appendix 5.1 A typical entry in the National Register of Basic Material.

```
┌─────────────────────────────────────────────┐
│              SELECTED SEEDS                   │
│                                               │
│   SPECIES:                                    │
│   IDENT NO (incl crop year)          ◯        │
│   QUANTITY:                                   │
│   REGION OF PROVENANCE:                       │
│   ORIGIN:                                     │
│                                               │
└─────────────────────────────────────────────┘
```

DETACH AND PLACE IN SACK	
FROM	
....	
....	
SPECIES	
IDENT NO.	YEAR (STAND NO)
CPT. NO.	
ADVICE NOTE (FORM SEED 2) NO.	

THIS IS SACK . . . . OF . . . . SACKS

FORM SEED 4
Rev'd 7/82

ATTACH SECURELY TO SACK	
FROM	
....	
....	
SPECIES	
IDENT NO.	YEAR (STAND NO)
CPT. NO.	
ADVICE NOTE (FORM SEED 2) NO.	

THIS IS SACK . . . . OF . . . . SACKS

Appendix 5.2 Labels suited for identifying cone
sacks and seed from selected seed
sources. (Labels are made from non-
tearable plastic and are coloured
green.)

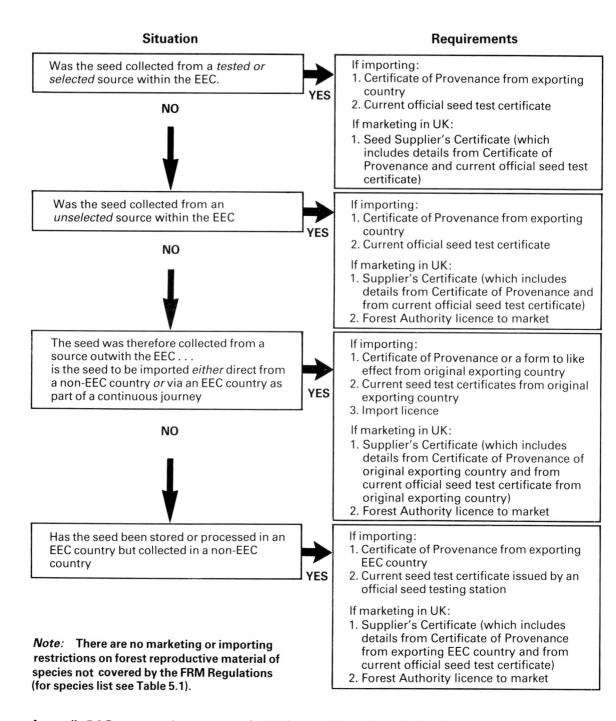

Situation	Requirements
Was the seed collected from a *tested or selected* source within the EEC. — **YES**	**If importing:** 1. Certificate of Provenance from exporting country 2. Current official seed test certificate **If marketing in UK:** 1. Seed Supplier's Certificate (which includes details from Certificate of Provenance and current official seed test certificate)
NO ↓ Was the seed collected from an *unselected* source within the EEC — **YES**	**If importing:** 1. Certificate of Provenance from exporting country 2. Current official seed test certificate **If marketing in UK:** 1. Supplier's Certificate (which includes details from Certificate of Provenance and from current official seed test certificate) 2. Forest Authority licence to market
NO ↓ The seed was therefore collected from a source outwith the EEC . . . is the seed to be imported *either* direct from a non-EEC country *or* via an EEC country as part of a continuous journey — **YES**	**If importing:** 1. Certificate of Provenance or a form to like effect from original exporting country 2. Current seed test certificates from original exporting country 3. Import licence **If marketing in UK:** 1. Supplier's Certificate (which includes details from Certificate of Provenance of original exporting country and from current official seed test certificate from original exporting country) 2. Forest Authority licence to market
NO ↓ Has the seed been stored or processed in an EEC country but collected in a non-EEC country — **YES**	**If importing:** 1. Certificate of Provenance from exporting EEC country 2. Current seed test certificate issued by an official seed testing station **If marketing in UK:** 1. Supplier's Certificate (which includes details from Certificate of Provenance from exporting EEC country and from current official seed test certificate) 2. Forest Authority licence to market

Note: **There are no marketing or importing restrictions on forest reproductive material of species not covered by the FRM Regulations (for species list see Table 5.1).**

Appendix 5.3 Documentation necessary for the importation and marketing of seed species covered by the Forest Reproductive Material Regulations 1977. For full legal details Statutory Instrument 1977 No. 891 should be consulted. For further explanation see sections of text entitled Marketing and Importation.

Chapter 6

Flower, Fruit and Seed Development and Morphology

*by **A. M. Fletcher***

Introduction

All concerned with the estimation and collection of seed crops require some basic knowledge of fruit or cone and seed development and morphology. Seed is produced by a complex series of biological events leading from the initiation of the flowering buds to their development, pollination, fertilization and finally to the seed which germinates. These events are normally completed in most coniferous (gymnosperm) and broadleaved (angiosperm) species over an 18-month period; pines require an additional year. There will be slight differences each year according to climatic variation, location, and positional differences within and between individual trees.

For the terms used in this chapter the reader is referred to the Glossary (p.122).

Conifers (gymnosperms)

In conifers there are two types of flowers (strobili); male or pollen producing cones and female or ovule producing cones. Conifers are monoecious, i.e. both male and female structures are produced separately on the same plant. In conifers, seeds develop (they are borne naked) on scales which are spirally arranged around a central axis to form a cone. Examples of the cycle of flower and cone development are given in Figures 6.1 and 6.2 for *Picea sitchensis* and *Pinus sylvestris* respectively.

Stages in cone development of *Picea sitchensis*

Formation and development of flower buds

Primordia of the reproductive or flower buds are initiated within vegetative buds and can only be seen under a microscope. A great deal of growth occurs within each vegetative bud before it swells and bursts open in late May or early June. Initiation of the bud scales of the terminal buds, which will eventually develop to become either vegetative or reproductive buds, commences in late March in the swelling bud from the previous growing season and continues until the completion of shoot elongation in early July. Axillary buds which might become reproductive buds develop while the shoots are elongating. During shoot elongation it is impossible to tell whether or not the primordia will eventually develop into vegetative, seed- or pollen-cone buds. In mid-July leaf initiation begins within the bud and there is a marked increase in cell division in the apex. This phase of very rapid activity lasts until mid-August and during it some buds undergo a transition to reproductive apices with the production of bracts or microsporophylls instead of leaves. The differentiation and development continues at a slower rate until the buds become dormant in early to mid-November. The buds then remain dormant until mid- to late March.

59

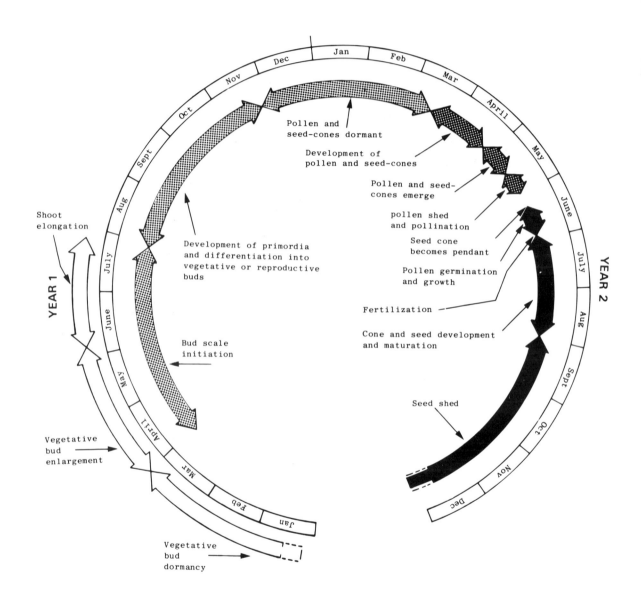

Figure 6.1 The cycle of flower and cone development in *Picea sitchensis*.

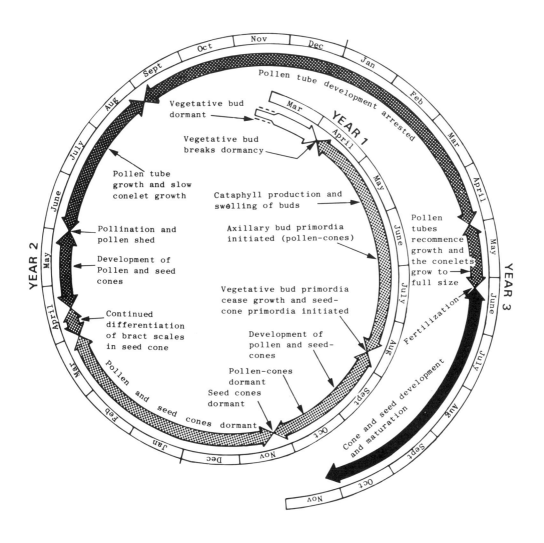

Figure 6.2 The cycle of flower and cone development in *Pinus sylvestris*.

61

Pollen-cones normally develop from small axillary or small terminal apices on slower-growing less-vigorous shoots more commonly in the lower portion of the green crown. Seed-cones develop from terminal apices on vigorous shoots or from vigorous axillary buds on these shoots in the upper crown. Before March in the year of flowering it is extremely difficult to identify flower buds from external appearances and accurate forecasts of developing cone crops cannot be made without detailed dissections of many buds.

Pollen-cone development

Pollen-cone buds break their winter dormancy in mid-March 4-6 weeks before pollen is shed. Each pollen-cone consists of a central axis which bears many microsporophylls in a spiral arrangement; each microsporophyll has two microsporangia (pollen sacs) on the vertical side, each of which is filled with many pollen mother cells (see Figure 6.3). Each of these

and in mid-May pollen is shed (see Plate 1). As the pollen-cone pushes through the bud scales it is normally reddish in colour and turns more yellow just before pollen is shed. When fully elongated, pollen-cones are 2-3 cm in length. The axillary positioned pollen-cones tend to grow upwards in the plane of the foliage and may develop a curved shape. Spent pollen-cones remain on the tree for several months; there is usually a long stalk at the base with the shrivelled brownish microsporophylls at the tip. Pollen collected for research purposes is harvested just as the first grains of pollen are shed and when moisture can no longer be squeezed out of the cones by finger and thumb pressure.

Seed-cone development

Each seed-cone consists of bracts, ovuliferous scales (megasporophylls) and ovules (megasporangia) forming a soft and fleshy flower (see Figure 6.4). There are two ovules at the base of each ovuliferous scale.

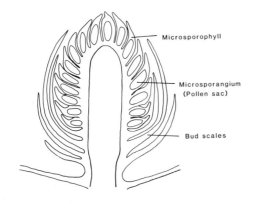

Figure 6.3 Median longitudinal section of a dormant pollen-cone of *Picea* spp. (x2).

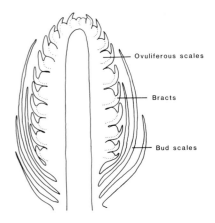

Figure 6.4 Median longitudinal section of a dormant seed-cone of *Picea* spp. (x2).

mother cells divides by meiosis (reduction division) to produce four pollen grains (microspores), each of which has half the number of original chromosomes (haploid) in the nucleus. This takes place in late-March after which the pollen-cone develops slowly and gradually swells. Towards the end of April the pollen-cones elongate and push through the bud scales; elongation then continues rapidly

Seed-cone buds break their winter dormancy towards the end of March and begin to elongate and swell immediately; any resin on the buds becomes clear and the buds develop a pointed shape and a very shiny appearance. Towards the end of April rapid elongation takes place and the seed-cone begins to push through the bud scales. They normally rupture near the middle or base of the cone to leave a small 'hat' of bud scales over the tip (see Plate 2) which is eventually removed by the wind. Seed-cones normally grow erect and as they emerge may vary in shade and colour between green and red. Maximum extension development takes place in the middle and lower portions of the seed-cone. Not all ovuliferous scales produce functional ovules (egg cells); those in the middle two-thirds of the cone have the greatest potential to produce functional ovules. The lowest scales and those nearest the tip never bear fertile seed.

Pollination and fertilization

Pollination and fertilization are two distinct events separated by an interval of 6-8 weeks. Pollination takes place during mid-May and fertilization in late June. Seed-cones are most receptive when the bracts separate and bend back to a horizontal position thus allowing pollen to enter between the bracts (see Plate 3). Pollen is transferred between trees by the wind and since the grains in some species have air sacs, pollen can be dispersed over a wide area if the climate is favourable at the time of release. The period of receptivity lasts for about one week after which the bracts close up, the seed-cone becomes 'woody' (see Plate 4) and finally turns downwards to a pendant position in early June.

At the time of pollination each ovule has within its wall a female gametophyte which, as in pollen formation, has a haploid number of chromosomes. The ovule has an elongated neck terminating in the stigmatic tip which is both sticky and hairy (see Figure 6.5). A pollination

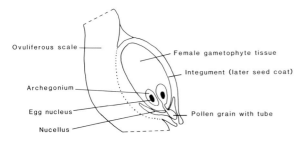

Figure 6.5 Diagram of a coniferous ovule at fertilization.

droplet is produced on the tip prior to pollination. Pollen lands on a bract scale, slides down the smooth upper surface of the bract to its base where the stigmatic tip is situated. The pollination droplet draws pollen into the micropyle; at the end of the receptive stage the tips of the stigma grow inwards and shrivel to enclose several pollen grains within the micropyle.

After pollination the female gametophyte develops and becomes mature and ready for fertilization by late-June. The pollen grains germinate about 2 weeks after pollination and after a further 3 weeks the pollen tubes grow to reach the archegonium where fertilization occurs, at which time chromosomes from the male parent are transferred to the female egg cell. A fertilized egg thus has the normal (diploid) complement of chromosomes restored.

Embryo and seed development

At the time of fertilization the seed-cone has developed almost to its final size although there are no external signs of embryo and seed development. Embryo development follows immediately after fertilization and by mid- to late July a club-shaped embryo is formed. Through a series of cell divisions the embryo develops a shoot apical meristem and root apex and then the shoot apex divides to form a ring of cotyledons. The embryo elongates and by the end of August it is almost the full length of the female gametophyte. A seed coat starts to differentiate at the time of fertilisation and is completely developed 6 weeks later. The seed wing, which is derived from the ovuliferous scale, starts to develop just after pollination and continues until late June. The seed coat encases the embryo which is surrounded by a milk-like substance (endosperm) containing stored food reserves for the embryo and germinating seedling.

As the seed-cone matures, it and the developing seed begin to dry, the seed endosperm becomes less milky and much firmer and the embryo grows to its full length. This happens about the beginning of September. The mature embryo which lies longitudinally in the seed consists of an upper part (mainly cotyledons) and a lower part (the hypocotyl) which includes the primary root (radicle). The radicle is covered with a root cap. At the apex of the hypocotyl there is a minute growing tip (epicotyl) which later develops into the plumule which itself later differentiates into stem and leaves (see Figure 6.6).

The ovuliferous scales turn from a green or reddish colour to brown and they become feathery and crisp to the touch (see Plates 5 and 6). As the drying continues the ovuliferous

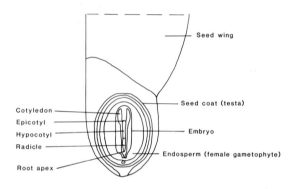

Figure 6.6 Diagrammatic representation of a mature seed of *Picea* spp.

scales begin to separate to allow seed to shed. The time of seed fall depends on the weather and varies from year to year. Seed fall normally begins in October and by the end of December most of the seed has gone although any remaining seed will continue to fall even until the end of February. The complete cycle of flower and cone development is summarised in Figure 6.1.

Cone development in *Pinus* spp.

The following section mainly applies to *Pinus sylvestris*, *P. contorta*, and *P. nigra* var. *maritima*, the three most commonly grown pines in Britain. *Pinus* spp. produce two types of shoots, long twig-like shoots and short-shoots from which needles develop. The pollen-cone and seed-cone buds are both differentiated within a long-shoot bud, seed-cones replacing future long-shoots and the pollen-cones future short-shoots.

Pollen-cones and seed-cones are normally borne on separate branches with the seed-cones in the upper crown on vigorous leading shoots. Pollen-cones normally occur in the lower crown on shoots which are less vigorous. In *Pinus contorta* both types of cones can often be found on the same shoot.

Pine shoots break their winter dormancy in early April with the production of sterile cataphylls which are found at the base of all shoots. In May the buds themselves begin to swell and elongate and further cataphylls are produced until mid-August at an increasing rate. Axillary bud primordia start to develop in the axils of these cataphylls at the time when half the total number of cataphylls have been initiated. The first of these, towards the base of the long-shoot, later become pollen-cones and those nearer the apex become vegetative short-shoots. To begin with there is little difference between the two types of primordia but by mid-August the vegetative primordia cease growth whilst the pollen-cones quickly grow; microsporophylls are formed spirally around the cone axis until early September. By October all the microsporophylls have been initiated and the pollen-cones are very distinct as a swollen region at the base of an upward turned long-shoot bud.

Seed-cone primordia are initiated in mid-August in the axils of the cataphylls just below the apex of the long-shoots. During September and October bract scale initiation begins but only those scales at the base of the cone axis are formed before the onset of dormancy in late October or early November. Because they are only partially developed at this stage it is usually impossible to tell the potential seed-cone crop for the following spring by external visual examination. Differentiation of bract scales continues after dormancy is broken in early April until completion in mid-April. Ovuliferous scales are developed and ovules form at the base on the upper (adaxial) surface; by late May or early June the seed-cone is ready for pollination (see Plates 9 and 10).

Pollen-cone buds recommence growth in early April and begin to enlarge in early May. Following meiosis they shed their pollen in late

Sitka spruce: *1. pollen-cones commencing shedding; 2. seed-cone: bursting bud; 3. seed-cone: receptive; 4. seed-cone: woody, post-pollination; 5. seed-cone: prior to collection; 6. seed-cone: mature.*

Lodgepole pine:

7. pollen-cone: developing;

8. pollen-cone: releasing pollen;

9. seed-cones: developing;

10. seed-cones: receptive;

11. seed-cones: woody, post-pollination;

12. seed-cones: prior to collection;

13. seed-cones: mature;

14. seed-cones: one and two-year-old cones.

 15

 16

Noble fir: *15. seed-cone: developing;*
16. seed-cone: woody, post-
pollination;
17. pollen-cones: developing.

 17

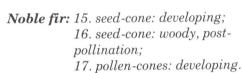 18

 19

European larch: *18. seed-cones:*
receptive;
19. seed-cone: mature;
20. seed-cones: new (left)
and old.

20

Japanese larch: top left
21. seed-cones: receptive;

Douglas fir:
top right & above left
22. pollen-cone: commencing shedding;
23. seed-cone: receptive.

Beech: above right
24. seed-cones: becoming woody; pollen-cone: shedding.

Birch: right & far right
25. pollen-cone: prior to shedding;
26. seed-cone: receptive.

Photo credits: Plates 1, 6, 7, 8, 11 and 19 are the copyright of Dr. R. Parks and are reproduced here by kind permission.
All other plates are by Dr. A. M. Fletcher, Forestry Commission.

May or early June. Pollen-cones, both when sheathed in bud scales and immediately after breaking through, are pale green in colour; they rapidly change to yellow at the time of meiosis (see Plates 7 and 8). Empty pollen-cones remain on the tree for 2-3 weeks before eventually falling to leave the lower portion of the long-shoot bare behind the zone of current year's needles; old 'male' shoots have a characteristic 'tufted' appearance for several years.

Pollination takes place in late May or early June. Some 2 weeks before this the seed-cones emerge from their bracts and are supported on a small stalk held in an erect position. Initially the seed-cones are pinkish in colour (see Plate 9); later these turn a deeper red. *P. sylvestris* seed-cones are 10 mm long; *P. contorta* seed-cones vary with origin and are from 10 to 15 mm long whereas *P. nigra* var. *maritima* seed-cones are larger still. As a seed-cone opens, the stalk supporting it holds it above the developing bracts and needles. The separation of fertile ovuliferous scales, which lasts for 3 or 4 days, begins near the base and progresses upwards at a rate dependent upon the weather (see Plate 10). Pollen grains which alight on the scales slide down towards the ovules at the base where a pollination droplet is produced which guides the pollen grains into the micropyle tube and the nucellus. As the ovuliferous scales close, the stalk begins to curve away from the shoot and the seed-cone finally becomes pendant (see Plate 11).

Pollen grains germinate within a few days of reaching the nucellus and the pollen tube develops very slowly during the first growing-season and development ceases temporarily two to three months after pollination. Pollen tubes resume growth in late April the following year and penetrate further into the nucellus. The female gametophyte also resumes cell division at the same time and several archegonia develop in each ovule. Fertilization takes place in early June, some 12–13 months after pollination. Embryo development proceeds until it is completed in late August for *P. contorta* (see Plates 12, 13 and 14) or in November for *P. sylvestris* and *P. nigra* var.

maritima, although the timing of events varies with latitude and season. The complete cycle and timings are summarised in Figure 6.2.

Cone development in *Abies* spp.

Abies grandis and *A. procera* (syn. *A. nobilis*) flush in mid- or late May and shoot elongation only lasts for 50–60 days. By mid-August vegetative buds are enlarged and externally appear to be fully developed. The vegetative buds however break dormancy 6–8 weeks before flushing. Bud differentiation takes place at the end of vegetative shoot elongation in early July and flower buds evolve from axillary buds. Pollen-cone and seed-cone buds are occasionally found on the same branch.

Pollen-cone buds are found on the underside of shoots in the middle or lower portions of the crown; they develop rapidly between the end of July and late October when they become dormant. At this time all the microsporophylls and microsporangia have been differentiated and the cones of *A. procera* are about 3 mm long; they are somewhat smaller in *A. grandis* and are found in clusters of up to 20. Buds of *A. procera* are reddish-brown while in *A. grandis* they have a brownish-green tinge; in both species they are covered in a heavy coating of greyish-white resin.

Seed-cones develop from adaxial axillary apices in the upper crown and commonly in the top eight whorls but in heavy flowering years or on isolated trees they may extend downwards in the crown. Bract initiation, which commences in late July or early August, with ovuliferous scale initiation starting about one month later, continues until the onset of bud dormancy in late October or early November. At this time the seed-cones are easily identified being 3–4 mm long in *A. grandis;* in *A. procera* they are larger and wider. In the former there can be up to six to eight seed-cones along each branch whereas in the latter there are normally only one or two (see Plates 15 and 16). They are normally found on different branches from those on which the pollen cones appear.

The pollen-cones break dormancy in early

March; by late April they are very swollen (see Plate 17) and shed their pollen in mid-May for *A. grandis* and 2–3 weeks later in *A. procera*. Spent flowers remain on the tree for several months.

The seed-cones of *A. grandis* enlarge in mid-April and elongate to burst through the bud scales in early May after which they continue to grow rapidly until pollination time. Development in *A. procera* is approximately 2 weeks later. *A. grandis* flowers are a yellowish-green and in *A. procera* greenish-purple with protruding green bract scales often tinged with red. The two ovules at the base of each ovuliferous scale are pollinated in mid-May or early June. The pollen germinates in early July and fertilization occurs in mid- to late July. Elongation of the seed-cone slows down during the period between pollination and fertilization but the diameter increases rapidly. After fertilization there is another burst of growth in length. Embryos and seed are mature in early September and the cones dry out rapidly. They break up completely and seed shedding begins in mid-September; normally by the end of October all that remains are the upright cones axes which stay on the trees for several years. As they mature cones of *A. procera* turn brownish in colour whereas in *A. grandis* the cones turn from a yellowish-green to a brownish-green. The seed set in most *Abies* spp. is often poor and frequently only 20 per cent of the seeds formed are viable.

Cone development in *Larix leptolepis* and *L. decidua*

Flower buds are differentiated at the end of the main period of long-shoot elongation towards the end of July. Both pollen-cone and seed-cone buds are normally differentiated on short-shoots at least one-year-old. Flower buds occur throughout the crown with pollen-cone buds most commonly on pendulous, less vigorous branches in the lower crown and seed-cone buds on the distal parts of more vigorous branches which often have an upswept habit.

Pollen-cone buds develop between July and the end of October and are fully developed before the onset of dormancy. In early November the pollen mother cells have partially completed meiosis which is resumed in the following February/March. These pollen mother cells can be seriously damaged by low temperatures during the early phases of meiosis or by severe low temperatures during the winter and this is often the reason for the production of infertile pollen. Seed-cones also develop during the July–October period but meiosis does not begin before the onset of bud dormancy.

The breaking of dormancy of flower buds, especially in *Larix leptolepis*, seems to be highly dependent on temperature and (depending on origin) can be weeks in advance of *L. decidua*. Pollen-cone buds rapidly enlarge during the completion of meiosis and emerge from their bud scales up to 3 weeks ahead of pollen shed. At this time the buds are rather flat and ovoid, often with a reddish tinge and they remain like this until the axis elongates rapidly. After the microsporophylls separate they turn a more definite yellow and shed their pollen, after which they turn a darkish-brown colour and may remain on the tree for several weeks.

The seed-cone elongates after dormancy and the shiny buds become very long and pointed before the bud scales burst. Depending on the season the buds may be in an obviously elongated state for 7–10 days before bursting. On bursting the seed-cones can vary in colour from red to green. Elongation after bud burst continues for another 2–3 weeks until pollination takes place, at which time the seed-cone is erect and continues to remain so (see Plates 18 and 19). The ovuliferous bracts and scales become reflexed and separate and the stigmatic tip remains receptive for several days. The cone continues to elongate during the pollination period and the ovuliferous scales thicken to fill the spaces between the bracts and so prevent further entry of pollen. Pollen grains swell and develop about 4 weeks after alighting on the stigma of the seed-cone; fertilization takes place 6–8 weeks thereafter. At the time of fertilization the ovule is almost fully enlarged. Embryo development proceeds until maturity in late September/October (see Plates 20-23). Often in *Larix* spp. there are many empty seeds in each cone due to poor pollination or embryo failure and in artificially made hybrid crosses especially, a high proportion of embryos are inviable.

Cone development in *Pseudotsuga menziesii*

In *Pseudotsuga menziesii* (Douglas fir) vegetative shoots flush in mid-May and shoot elongation ceases in early July. Lateral buds which could eventually develop into pollen- or seed-cones are initiated before bud-burst in April; differentiation takes place several weeks later. By October, when the buds become dormant, the different bud types can be identified externally. Pollen-cone buds occur in the leaf axils along the entire length of the shoot but more often in the middle or near the base of the shoots. The pollen-cones are a light orange-brown colour and have fewer bud scales than vegetative buds. Seed-cones, which occur near the ends of the shoot, are longer and broader than the pollen-cone or vegetative buds.

Pollen-cone buds start growing in early March and by late March the buds begin to swell and burst through the bud scales in mid-April. The pollen-cone, which is a pinkish-red colour elongates, turns downwards, develops a more pinkish-yellow colour and is about 2 cm long when fully developed (see Plate 24). It sheds pollen at the end of April and falls from the tree within a few days.

Seed-cones also start growth in early March. The cone axis elongates and finally bursts through the bud scales to assume an upright position in mid-April. Seed-cones are 3 cm long and are normally a reddish colour but green coloured flowers can also be found. The seed-cone has very long extended bracts which are reflexed right back at the time of pollination in late April. They remain in the receptive stage for 4 or 5 days (see Plate 25).

Seed-cones turn into a fully pendant position almost immediately after pollination; they elongate rapidly and reach their final size in

early July. Fertilization takes place in mid- to late June. Each ovuliferous scale has two ovules at its base. The embryo develops for about 10 weeks after fertilization before reaching maturity in mid-September. The cone is greenish in colour during its growth phase but as it matures and dries in late August the bracts turn a light brown. Seed fall can begin towards the end of September and the majority of the seed has fallen by early November.

Broadleaved trees (angiosperms)

Each genus of broadleaved tree exhibits its own peculiarities of flower form and behaviour. Staminate (male) and pistillate (female) flowers may be borne on different trees as usually occurs in *Fraxinus excelsior* or separately on the same tree as in *Quercus* spp., *Fagus* spp. and *Betula* spp. Alternatively the male and female organs may occur in the same flower as in *Aesculus* spp. In cases where wind is the main vector for pollen transport the pistillate flowers are normally small and drab with much larger pollen producing catkins, e.g. *Quercus* spp. Large-petalled colourful flowers are commonly found amongst the insect-pollinated species, e.g. *Sorbus* spp. and *Prunus* spp. and special mechanisms are needed for those species which are naturally self-pollinating, e.g. *Laburnum* spp. One kind of angiosperm flower and its main parts is illustrated in Figure 6.7. A fuller description of the different

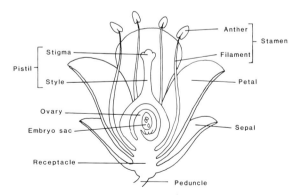

Figure 6.7 Diagram of a complete angiosperm flower.

types of flower formation is given in Forestry Commission Bulletin 59 *Seed manual for ornamental trees and shrubs.*

Flower development

In angiosperms the pollen grains are produced in an anther, a sac-like part of the stamen. The pollen mother cells undergo division similar to that in the conifers but each contains two sperm-cells.

Pollination occurs when pollen grains land on the stigma of the pistil which is normally coated with hairs, glands or a sticky fluid to catch them. Shortly after landing on the stigma, viable pollen grains germinate and produce a thin tube which grows down inside the style towards the ovule (see Figure 6.8). Growth of the pollen tube with its two sperm-nuclei in the tip is rapid and the time from pollination to fertilization is usually 1–2 days.

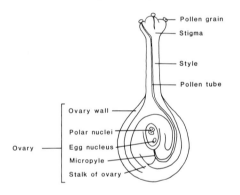

Figure 6.8 Diagram of an angiosperm pistil at fertilization.

The ovule of the angiosperms is more complex than that of gymnosperms. Amongst other organs it contains a mature embryo sac and an egg cell. When the pollen grain passes through the micropyle and fertilization takes place, one of the sperm-cells fuses with the egg nucleus to form a zygote which later develops into the seed embryo. The other sperm-cell fuses with two other nuclei and leads to the development of the seed endosperm. After fertilization the development of the endosperm proceeds rapidly; the development of the embryo follows later. In species such as Quercus and Acer the embryo becomes very large and the endosperm is consumed during its development. The food reserves for the germinating seedlings in such species are stored in the thick fleshy cotyledons. A seed coat (testa) develops to enclose the embryo which is fully mature in most species by the time seed fall occurs.

Quercus spp.

Quercus spp. are monoecious and flowering coincides with early leaf development in April or May. The staminate flowers occur in long, slender, pendulous clustered catkins arising in the axils of leaves of the previous year or from terminal buds. Each flower has six stamens with bright yellow anthers. The pistillate flowers occur in the axils of current year's leaves with normally a three-celled ovary and a similar number of pistils; as a consequence of abortion only one seed matures. The seed is marked at the base by the abortive ovules and partially enclosed by a scaly cup. In *Q. petraea* the pistillate flowers are in groups of two or three on a very short stalk (2–7 mm); the acorns subsequently have very short stalks. In *Q. robur* the acorns are carried on long stalks or peduncles (13–25 mm). The acorns are green when developing and as they ripen they turn a brown colour and begin to fall in October, most heavily after a keen air-frost.

Fagus sylvatica

Fagus sylvatica is monoecious and flowering occurs in May. The staminate flowers have 8 to 16 stamens and greenish anthers and are in long-stemmed globular clusters at the base of the current year's shoots or in the axils of the lowest leaves. The pistillate flowers are found mainly in the axils of the distal leaves and are normally in pairs with three styles and two ovules in each (see Plate 26). The pistillate flower is surrounded by numerous oval-shaped hairy bracts which form a four lobed involucre which later becomes woody and covered with recurved prickles (husk); it encloses the two one-seeded nuts. The stalk of the husk is downy. The husk turns brown as the nuts mature and by late September or after a keen frost the husk splits to release the nuts. Earlier falls of seed mainly consist of empty shells.

Betula pendula

Betula pendula is monoecious and flowering occurs in late April or early May. The staminate catkins are formed in August and remain naked throughout the winter. They are found either singly or in clusters in the axils of the last leaves of a long shoot or on the ends of short lateral branches. In spring the staminate catkins elongate rapidly to two or three times their dormant length (approximately 3–6 cm) and the scales are dark chestnut-brown in colour, sometimes tinged with the yellow of the anthers (see Plate 27). The catkins are cast after pollen has been shed. Each individual flower has two stamens which are split into two minutely-stalked anthers. The pistillate flowers, which are cone-like and borne terminally and erect on short shoots, elongate as the leaves begin to emerge (see Plate 28). Initially they are erect, greenish-yellow in colour but sometimes have a reddish tinge and are smaller (2–3 cm) than the staminate flowers. Each flower has two styles and the ovary is two-celled, although only one cell produces a fertile seed. As the seed-cone matures it becomes pendulous and brownish in colour and seed fall occurs from late August onwards.

REFERENCES and FURTHER READING

ALLEN, G.S. and OWENS, J.N. (1972). *The life history of Douglas fir.* Environment Canada, Canadian Forestry Service.

BRAMLETT, D., BELCHER, E.W., Jr., DE BARR, G.L., HERTEL, G.D., KARRFALT, R.P., LANTZ, C.W., MILLER, T., WARE, K.D., and YATES, H.O. III. (1977). *Cone analysis of southern pines. A guide book.* USDA Forest Service General Technical Report SE-13.

DOBBS, R.C., EDWARDS, D.G.W., KONISHI, J. and WALLINGER, D.P. (1976). *Guideline to collecting cones of BC conifers.* British Columbia Forest Service Canadian Forestry Service, Joint Report No. 3.

EIS, S. and CRAIGDALLIE, D. (1981). *Reproduction of conifers. A handbook for cone crop assessments.* Canadian Forestry Service, Report BC-X-219.

GORDON, A.G. and ROWE, D.C.F. (1982). *Seed manual for ornamental trees and shrubs.* Forestry Commission Bulletin 59. HMSO, London.

KRUGMAN, S.L., STEIN, W.I. and SCHMITT, D.M. (1974). Seed biology. Chapter 1 in *Seeds of woody plants in the United States.* USDA Forest Service Agricultural Handbook 450, 5–40.

MOIR, R.B. and FOX, D.P. (1975). Bud differentiation in Sitka spruce, *Picea sitchensis* (Bong.) Carr. *Silvae Genetica* **24**, 193–196.

OWENS, J.N. and MOLDER, M. (1976a). Bud development in Sitka spruce. I. Annual growth cycle of the vegetative buds and shoots. *Canadian Journal of Botany* **54**, 313–325.

OWENS, J.N. and MOLDER, M. (1976b). Bud development in Sitka spruce. II. Cone differentiation and early development. *Canadian Journal of Botany* **54**, 766–779.

OWENS, J.N. and MOLDER, M. (1980). Sexual reproduction in Sitka spruce (*Picea sitchensis*). *Canadian Journal of Botany* **58**, 886–901.

OWENS, J.N. and BLAKE, M.D. (1984). The pollination mechanism of Sitka spruce (*Picea sitchensis*). *Canadian Journal of Botany* **62**, 1136–1148.

OWENS, J.N. and MOLDER, M. (1984). *The reproductive cycle of lodgepole pine.* Ministry of Forests, Province of British Columbia, Canada.

SARVAS, R. (1962). Investigations on the flowering and seed crop of *Pinus sylvestris*. *Communicationes Instituti Forestalis Fennica* **53** (4), 1–198.

TOMPSETT, P.B. (1978). Studies of growth and flowering in *Picea sitchensis* (Bong.) Carr. 2. Initiation and development of male, female and vegetative buds. *Annals of Botany (London)* **42,** 889–900.

WAREING, P.F. (1958). Reproductive development in *Pinus sylvestris*. In *The physiology of forest trees,* ed. K.V. Thimann, 643–654. The Ronald Press Co., New York.

Chapter 7

Identification and Assessment of Cone and Seed Crops

by **A. G. Gordon** *and* **R. Faulkner**

Introduction

The seed crop of a forest is a valuable resource and should be regarded as a potential additional source of income to the owner. As natural forests in other parts of the world have been cut so the amount of seed available for export has fallen and the costs risen. Seed crops in plantations far removed from their natural habitat have thus often become extremely valuable additional sources of seed of particular origins.

In Britain as more forests have reached seed-bearing age, their potential to produce large seed crops has increased. Clear evidence of this has been seen in the record harvest from *Picea sitchensis* in 1983. This followed a long period in the 1960s and 1970s when the seed production of British forests was disappointingly low (see *Report on Forest Research* for the years 1960 to 1981).

A crop of forest tree seed cannot be successfully harvested without detailed planning. The earlier the warning and the earlier that the potential of the crop can be assessed the sooner these plans can be made. Crop assessment is not easy and can only be done with any degree of accuracy by experienced staff.

Seed collections by private collectors are possible from Forestry Commission forests, provided the Forestry Commission itself has no prior plans to make collections from the stands.

Early prediction of seed crops

The earliest hint of a potential cone crop can be gained in some conifer species such as *Pseudotsuga menziesii* and *Abies* spp. by the formation of flowering buds in the late summer of the year preceding the cone harvest. These buds cannot be observed from the ground but can be readily seen on the tips of branches on trees felled or blown in the winter. Although they are not positive predictors of a crop, they nevertheless give an early indication of the possibility of a crop.

It is widely reported that a long, warm, dry summer is often followed by a good seed year. There is undoubtedly some truth in the report but the relationship is by no means perfect.

It is not until the spring flush of foliage that it is possible, in most species, to begin to assess the true extent of the potential crop. There can be no seed crop without male and female flowers but the presence of abundant flowers regrettably does not guarantee an economically collectable seed crop since late frosts and the particular weather conditions during flower and cone development play an overriding influence on the final crop (see next section).

All conifers are monoecious and in some species both male and female flowers are easily distinguishable (*Pinus* spp.) but in others (e.g. *Picea* spp. and *Pseudotsuga menziesii*) young female flower buds in late May bear remarkable similarity to vegetative buds. Illustrations of male and female flowers of the important conifer species have been included in Chapter 6 (see colour section).

In *Fagus sylvatica* the different flowers are almost indistinguishable without close examination. In *Quercus* spp. the presence of the feathery male flower catkin can easily be spotted from a distance just before full leaf expansion, but the female flowers can only be identified by close examination. Winter catkins of *Betula* spp. are easily seen indicators of male flowers throughout the winter months but the female catkin is less easily seen after leaf flush.

Early warning of a heavy crop may sometimes be gained from surprisingly great distances. In years of exceptional flowering in some conifers, the crowns from a distance may appear yellow or orange-brown in colour due to the massive production of male flowers. However, the presence of abundant male flowers does not always indicate abundant female flowering.

Anyone interested in collecting seed should keep a notebook handy at all times for recording.

Weather conditions play an important part in the ability of an observer to see a potential crop. Dull weather with no light and shade contrasts can completely hide tell-tale signs of flowering and the impressions gained when looking towards the sun often bear little resemblance to those of the same flowering when seen from the opposite direction.

A summary of flowering and fruiting data for major conifers and broadleaves is given in Table 7.1. Actual flowering periods vary from year to year depending upon the lateness or earliness of spring. Flowering in northern and upland Britain tends to be 2-3 weeks later than that in the south and lowlands.

It is important to assess the crop prospects regularly during the development of the fruits. To do this a good sample of fruits should be

Table 7.1 Crop forecast data for seed of commercial forest trees.
(Adapted from Matthews, 1955; Seal, Matthews and Wheeler, 1963; and Technical Note No. 48, CEMAGREF, Ministry of Agriculture, France – for full address, see page 36).
Note: Much of the information in this Table is as recorded in the French publication – it should be borne in mind that data for seed crops grown in Britain may vary from the figures quoted.

Conifers

Characteristic		Abies alba	Abies grandis	Abies procera	Chamaecyparis lawsoniana	Larix decidua
Age of first good seed crop (years)		30	40	20	20–25	15–20
Age of maximum production (years)		40–60	45–50	40–60	40–60	40–60
Periodicity of fruiting (years)		2	3–5*	2–4	2–3	3–4
Period for crop assessment		Mid June	End of May	Mid June	May–June	May–June
Timing of cone collection	*earliest*	Early Sept	Mid Aug	End of Aug	Mid Aug	End of Oct
	normal	Late Sept–mid Oct	Early–mid Sept.	Early–mid Sept	Early–mid Sept	Dec–Jan
	latest	Late Oct	Late Sept	Mid–Oct	Mid–Sept	Feb
Collectable mature seed bearing trees per hectare	*average crop*	20	30	20	20	20
	heavy crop	60	60	60	60	80
Volume of cones per tree (hl)	*average crop*	0.2–0.7	0.2–0.7	0.2–1.5	0.2–0.8	0.1–0.5
	heavy crop	0.3–0.8	0.3–0.9	0.3–3	0.4–1.0	0.3–0.8
Volume (hl) of cones per hectare	*average crop*	4–15	6–20	6–30	10–20	2–10
	heavy crop	18–50	25–70	25–100	20–40	25–65
Yield of seed (kg) from one hectare	*average crop*	12–75	18–110	20–150	?	1.8–25
	heavy crop	55–250	75–385	100–500	?	22–155
Collection potential (hl per man day)	*felled trees*	8–12	8–12	10–15	1–3	2–4
	climbing	1.5–5	1.5–5	2–8	1–3	1–2

*Abies grandis does not flower as regularly as A. procera in UK.

Conifers *continued*

Characteristic		*Larix leptolepis*	*Larix x eurolepis*	*Picea abies*	*Picea sitchensis*	*Pinus contorta*
Age of first good seed crop (years)		15–20	15–20	30–35	30–35	15–20
Age of maximum production (years)		40–60	40–60	50–60	40–50	30–40
Periodicity of fruiting (years)		3–4	3–4	4–5	3–5	3–5
Period for crop assessment		April–May	May	End of June	End of May	June–July
Timing of seed collection	*earliest*	Sept	End of Sept	Early Oct	Mid Sept	Early Sept
	normal	Oct	Oct	Oct	Oct	Oct
	latest	Nov	Nov	Late Nov	Late Oct	Mid Nov
Collectable mature seed bearing trees per hectare	*average crop*	20	20	20	30	30
	heavy crop	80	80	60	60	60
Volume of cones per tree (hl)	*average crop*	0.1–0.5	0.1–0.5	0.3–0.7	0.3–0.8	0.01–0.05
	heavy crop	0.3–0.8	0.3–0.8	0.3–0.9	0.5–1.2	0.03–0.08
Volume (hl) of cones per hectare	*average crop*	2–10	2–10	6–15	6–15	0.2–1.0
	heavy crop	25–65	25–65	18–55	30–70	2.0–4.0
Yield of seed (kg) from one hectare	*average crop*	1.8–12	1.8–20	6–25	5–20	1–10
	heavy crop	22–80	22–100	18–85	25–100	10–30
Collection potential (hl per man day)	*felled trees*	2–4	2–4	8–10	3–4	1–3
	climbing	1–2	1–2	2–3	0.5–1.5	0.5–2

Characteristic		*Pinus nigra* var. *maritima*	*Pinus sylvestris*	*Pseudotsuga menziesii*	*Thuja plicata*	*Tsuga heterophylla*
Age of first good seed crop (years)		25–30	15–20	30–35	20–25	30–35
Age of maximum production (years)		60–90	60–100	50–60	40–60	40–60
Periodicity of fruiting (years)		3–4	2–3	4–6	2–3	3–5
Period for crop assessment		June–July	June–July	Mid May–mid June	May–June	June
Timing of seed collection	*earliest*	Late Nov	Mid Nov	End of Aug	End of Aug	End of Aug
	normal	Dec–Feb	Dec–Mar	Sept	Sept	Sept
	latest	Mar	Early Apr	Early Oct	Early Oct	Early Oct
Collectable mature seed bearing trees per hectare	*average crop*	30	30	10	20	20
	heavy crop	60	60	50	60	60
Volume of cones per tree (hl)	*average crop*	0.1–0.6	0.01–0.08	0.05–0.4	0.2–1.5	0.2–1.0
	heavy crop	0.3–1.0	0.05–0.15	0.1–0.8	0.5–3.0	0.5–2.0
Volume (hl) of cones per hectare	*average crop*	3–18	0.3–2.4	0.5–4	10–30	2–5
	heavy crop	18–60	3–7.2	5–25	30–50	10–15
Yield of seed (kg) from one hectare	*average crop*	2.5–22	0.18–2.8	0.2–4	?	?
	heavy crop	15–90	1.8–8.6	2–25	?	?
Collection potential (hl per man day)	*felled trees*	2–5	0.8–1.5	3–6	2–4	2–4
	climbing	1–3	0.2–0.5	2–3	1–3	1–3

Notes: *Pinus* spp. cones develop over 1½ years.
Some idea of potential crop in year 2 can be made after the first year.
Collection dates refer to year 2.

Characteristic		Betula spp.	Fagus sylvatica	Quercus petraea	Quercus robur	Quercus rubra
Age of first good seed crop (years)		15	50–60	40–50	40–50	30–40
Age of maximum production (years)		20–30	80–200	80–200	80–120	80–120
Periodicity of fruiting (years)		1–2	2–8	3–8	2–6	2–3
Period for crop assessment		July	July–Aug	Aug	Aug	Aug
Timing of seed collection	*earliest*	Aug	Early Oct	Early Oct	Early Oct	Late Aug
	normal	Sept	Oct–Nov	Oct	Oct	Sept
	latest	Oct	Late Nov	Early Nov	Early Nov	Oct
Collectable mature seed bearing trees per hectare	*average crop*	20	20	20	15	15
	heavy crop	60	40	40	30	30
Weight of seed per tree (kg)						
	average crop	1–3	8–10	20–40	30–50	30–50
	heavy crop	3–6	15–20	70–90	110–130	110–130
Weight of seed per hectare (kg)						
	average crop	3–10	150–200	500–700	500–700	500–700
	heavy crop	6–20	400–600	<3200	<3600	<3600
Collection potential (kg per man day)						
	from ground	2–5	1	50	50	50
	by vacuum	—	3	—	—	—
Weight (g) of one litre of fruit		—	400–500	500–700	600–800	500–700

obtained. Since fruits near ground level on edge trees are often poorly developed due to poor pollination and fall early, it is important to obtain fruit from the upper crown of the trees; these can be obtained using a shotgun.

Factors affecting cone and seed crops and collection targets

Crop forecasting

It is not easy to make an accurate assessment of the seed potential of a particular forest stand. Many factors affect the yield, weather conditions at the time of pollination having a particularly important influence. Continuous heavy rain throughout the period when female flowers are receptive reduces the number of successful cross pollinations and consequent fertilizations, resulting in a low seed yield per cone. Similarly, an air frost at just the wrong time can have a devastating effect, although the differences in the susceptibility of individual trees in a stand, due to stage of flower development, aspect, altitude and micro-location (hillock or hollow) can be dramatic.

Collection

Conifer cones are most readily collected from felled trees. Seed collectors are able to harvest crops from felled trees provided the desired harvest date coincides with the felling date. Seed collection is made from the floor beneath standing crops of *Fagus sylvatica* and *Quercus* spp., and the need to co-ordinate with felling schedules does not arise. However, with *Betula* and other small-seeded species, collection is made easier if the fruiting trees are felled.

Seed crops on *Fagus sylvatica* and *Quercus* spp. in any one year can vary widely between different stands of the same species growing only a few miles apart. The only certain method of seeing if there is a collectable crop present is to visit each stand. Most registered seed stands of *Fagus sylvatica* and *Quercus* spp. are on private estates.

The assessment procedure for conifers

Whether a visit is made to a stand simply to

look for cones or to attempt to assess accurately the quantity of seed available, it is crucial to choose the right weather conditions. In overcast conditions with no contrasts of light, cones (even when mature and contrasting in colour with the foliage) can be virtually impossible to spot even through powerful binoculars. Full overhead light can also cause strong shadows which hide the cones lower on the crown. The best time is either early or late in the day when the sun is low, and the best position is with the sun behind the observer.

In order to assess the potential of a whole stand it is necessary to be able to examine all of it. This is not always easy in practice but in undulating terrain it is usually possible to find a vantage point at a convenient distance away. In other stands, where the ground is flat or the slope even, the crops have to be judged by looking at the edge trees, by looking through gaps in the canopy and sometimes even by examining the forest floor for tell-tale signs of spent male and female cones broken off in strong winds. However, care should be taken not to give undue weight to roadside trees which have bigger crowns and often flower more heavily.

The first step is to make a general subjective assessment of the cone crop on the stand. This is done by looking through ×6 or ×8 binoculars from the best position available and trying to estimate how many cones there are on a given proportion of the trees. Obviously the greater the number of trees observed the greater the reliability of the information. However, it is extremely difficult to put into words and numbers what constitutes light, average, and heavy cone crops but the following classification is an attempt.

Absent	No cones on any trees.
Light	A few cones on about one tree in every 50.
Moderate	A significant number of cones visible in about 25–50 per cent of the trees.
Heavy	Very many cones (so many that it is difficult to count them) on

some (5–10 per cent) trees, a significant number of cones on many other trees and at least a few cones on nearly every other one.

The number of cones that constitute light, moderate and heavy classes unfortunately varies with species. The heaviest coning in *Abies* spp. may mean 100 to 200 cones per tree, whereas in *Picea sitchensis* it may mean 1000 or even, in exceptional circumstances, 2000 visible cones per tree. *Picea* spp. and *Abies* spp. tend to carry all their cones in the upper six branch whorls, whereas *Pinus* spp. and *Larix* spp. carry their cones all over the crown. Some pines, e.g. *Pinus nigra* var. *maritima,* often hide many of their cones behind the end tuft of needles and are difficult to spot even from very close range.

When assessing cone crops it is very important to eliminate the previous season's cones from the estimates. For *Pinus* spp. this is relatively easy at all seasons but for *Picea* spp. and *Larix* spp. the nearer to current crop maturity the assessment is made the more difficult it is to differentiate at a distance between current and previous year's cones.

In order to quantify the actual volume of cones in a particular stand the assessor should try to count with the aid of binoculars the actual number of cones seen on at least 20 randomly selected dominant trees in the stand. This can be multiplied by 4 in orchards, by 3 in ride-side trees and by 2 for trees in the interior of a stand, to give an estimate of the number of cones on both sides of the tree. By multiplying the average number of cones per tree by the estimated number of similar trees within the stand, an estimate of the total number of cones can be gained. The yield of the stand is estimated by dividing the number of cones by the average number of cones in one hectolitre (see Table 7.1). However, such quantitative assessments are difficult to perform and can never be relied on to be precise. Collections as much as four times more than the volume estimated have been achieved occasionally in practice and illustrate the potential variation involved.

The volume occupied by a certain number of cones varies greatly because there is much between-tree variation in cone size. In *Picea sitchensis* and *Pinus sylvestris* this variation in size can be as much as two-fold. This may dramatically affect the estimated volume of cones assessed; it may not in practice affect the weight of seed so much. As a rough yard-stick, it should be possible to collect 50 hectolitres of cones of *Picea sitchensis* from one hectare of mature trees coning moderately heavily, up to 100 hectolitres from one hectare in a very heavy crop; up to 25 hectolitres per hectare in a heavy crop on a pole-stage plantation. (One hectolitre is 100 litres or 2.7 imperial bushels. One half hectolitre of cones in a sack does not quite fill a 50 kg grain sack but it is rather more than a 25 kg potato sack can accommodate.)

Factors affecting yield of conifer seeds

When attempting to extrapolate the expected yield of seed from the volume of cones estimated, there are several variables which determine the result. The main ones are: the percentage of seeds which have developed normally; and the size and weight of the seeds. Conditions at the time of pollination determine to a very great extent the number of normally developed, or so-called full seeds that are formed. If conditions are dry, sunny and with some wind, pollination will be good and almost all seeds in a cone will be full. If conditions are dull and wet throughout the pollination period, poor pollination will take place and fewer full seeds will be formed. This is normally the result of self-pollination (pollination of the female cones by pollen from the same tree) which causes the embryo to abort during development, resulting in a fully developed seed coat which is hollow beneath. Occasionally seeds derived from self-fertilization do reach full maturity but the resulting seedlings are usually less vigorous than plants derived from seed from cross-fertilized flowers.

In order to assess the number of full seeds in the crop a random sample of at least five cones taken from the mid to upper crown of each of at

Figure 7.1 New Canadian designed cone cutter (F.D. Barnard, P.O. Box 144, Blind Bay, B.C., Canada, VOE 1HO). (*Photo: John Poole Photography*).

least 20 trees should be cut in half with a sharp blade or guillotine (see Figure 7.1) and the number of full and empty seeds in the surface of one half of the cones counted. Numbers of full seeds per cut surface above 3–4 for *Pinus sylvestris*, 6–8 for *P. contorta*, 8–10 for *Picea sitchensis*, 5–6 for *Pseudotsuga menziesii* and 2–3 for *Larix* spp. indicate collectable crops; only if seed is scarce and urgently needed is it worthwhile collecting crops with lower seed counts per cut surface.

Yields will obviously be lowered if cones are collected after seed has started to be shed. For *Pinus contorta, Picea sitchensis* and *Pseudotsuga menziesii* the period between full cone and seed maturity and natural seed shed is relatively short. A pause in cone collection of one week in these species may result in significant reductions in yield due to natural seed fall. In other species where the interval between full cone maturity and the onset of seed shedding is more extended, yields will not be affected by relatively short pauses unless the start of collection is very delayed. As an example of the effect of year to year variation, in both 1984 and 1985 the safe date by which Sitka spruce cones had to be collected was some 3–4 weeks earlier (mid-October 1985, end of October 1984) in Scotland than in England and

Wales where in 1984 economic collections were made until the very end of November from standing trees.

It is often wise to make full seed counts of cones collected from young seed orchards during the first few years after the onset of flowering when the pollen cloud is light or poorly distributed. Low pollen production results in a low set of full seeds which may render the collection uneconomical.

The size and weight of seed also varies between individuals, and the average weight of seeds will additionally be affected by weather conditions and latitude to some extent. In a very warm and dry year seeds may not grow as large or be filled so completely as in normal years. In the same way, seed of *Pinus sylvestris* from northern Britain has been found to be smaller than seed from the same species grown in the south and seed from seed orchards has tended to be heavier than from stands. Differences in cone size and shape and seed size are particularly easy to see between the different clones in a seed orchard, but experience has shown that the yield of seed per unit volume of cones is not directly related to cone size. Table 9.1 provides information on the range and average number of cones per hectolitre which can be used to convert estimates to volumetric units and also for estimating the final yield of cleaned seed per hectolitre of cones.

The assessment procedure for broadleaves

The task of assessing the potential crop of broadleaves is even more difficult than for conifers.

Acorns of *Quercus* species only develop sufficiently to be seen, without really careful examination, in late August and September. The presence of flowers in spring, as with conifers, is by no means a reliable indicator of a seed crop. Frosts can be particularly devastating to *Quercus* spp., as they tend to flower before full leaf extension and at a time when frosts are still not uncommon. It is a fairly common sight to see a clear line between the bottom badly frost-burnt half of a mature oak growing in a hollow, and the top which is untouched. By the time acorns have reached full maturity in late September to October it is usually possible by examination of the crown from a suitable vantage point, even without binoculars, to see if there is a worthwhile mast. A really heavy crop of acorns is one in which from about 30 metres away the crown of the tree appears to have acorns at the end of almost every twig. A light crop is one in which, from the same distance, it is possible to contemplate counting the individual acorns. Most often fruiting is intermediate between these two descriptions. From a tree with a large crown bearing a heavy crop it is quite normal to collect over 100 kg of acorns. A light crop could be as low as 5 kg per tree. Care must be taken to distinguish crops attacked by the knopper gall wasp (see Insect attacks).

The flowering and fruiting characteristics of the true native *Quercus* spp. are described in Chapter 6. The descriptions given are those of the pure species. Collectors must, however, be aware that in many localities great care will be necessary in identifying *Quercus* species because of the occurrence of hybrid or intermediate forms, and the presence of Turkey oak which often flowers heavily.

Flowering in *Fagus sylvatica* is not so easy to see from a distance as it is in *Quercus* spp. The flowers appear at about the same time as the first leaves are fully opened and from a distance tend to be hidden. As the male flowers mature and dry they become more obvious and are very easily seen after they fall to the forest floor. From close-up the feathery male and female flowers are easily distinguished. Later, after pollination, the female flower develops rapidly and soon the typical fruit cupules can be seen clearly from a distance. Even in high forest conditions flowering in beech can be identified at this stage because of the discarded stamen-bearing male flowers littering the forest floor.

The fruit of *Fagus* reaches nearly its full size within 2–3 weeks of pollination. At this time the fleshy valves of the cupule form the major part of the fruit and it is possible to tell if the fruit is fertile; the fleshy embryo will be clearly seen crammed into the triangular space formed by the valves of the cupule. The silvery walls of the ovule continue to grow even if the embryo does not develop. They can be mistaken for the embryo at first but once the creamy white flesh of the embryo is seen in a fertile seed, mistaken assessment should never occur again. As the embryo grows, the walls of the cupule become increasingly woody and the fruit becomes difficult to open. The embryo, seed coat and fruit walls are then best examined by cutting midway across the fruit. As the fruits mature they change from a green to a reddy-brown colour and in a heavy crop the boughs noticeably bend under the weight of the developing seeds.

As with *Quercus* spp., it is virtually impossible to gain an accurate idea of the size of the crop. It is impossible to count the number of fruits and not all fruits will be full. Experience, coupled with the ratio of full to empty seeds observed in the samples, will enable a sensible guess of the potential crop to be made. But crops vary due to the size of the seeds which are certainly influenced by the weather during development. Individual seeds in the 1984 crop were about 25 per cent smaller than beech seed imported from the continent.

In *Betula* spp. the potential of a crop is clearly seen all through the winter. Male catkins form in the late autumn on the tips of twigs and remain unopened until the following spring. The female catkins form after the leaves open and are upright, initially smaller, green and much more difficult to spot. Female catkins are seldom, if ever, found on trees that do not have male catkins. The size of the crop is again difficult to judge until the female catkins have matured and have assumed their typical pendulous postion. Then it is possible to categorise the crops as light, medium or heavy. It is normally not worth harvesting any but heavy cropping trees. As a rough estimate a fully grown tree with full crown and a heavy crop should produce about 10 kg of seed.

Details of seed crop estimation for gean, ash and sycamore can be found in Forestry Commission Bulletin 59 (Gordon and Rowe, 1982).

Factors affecting yield of broadleaved seeds

The actual crop collected from a stand of *Quercus* can vary greatly due to prevailing conditions. A warm and wet autumn will cause the acorns to be retained, a dry autumn with frosts and strong winds will tend to accelerate acorn drop. In the former the fallen acorns are more at risk to predators as it is normal to wait for some time between collection visits. In the latter the majority of acorns can fall in a few days and be harvested at one visit. Weather conditions at harvest have caused wide divergences between forecast crops and actual yields in recent years. An acorn crop is only sure when it is in the bag!

The presence of flowers does not always mean that good seed will be produced. While frost seldom seems to affect the flowers, the weather conditions at pollination are, however, critical and sunny, breezy weather is optimal. Even after successful pollination the crop is not always assured, as for example in 1976 when very heavy flowering and perfect pollination conditions were followed by crop failure all over Britain. The very intense late spring and summer drought seemed to result in the abortion of developing fruit.

In *Betula* spp. severe frosts at the time of male catkin extension can greatly affect the amount of fertile seed in the female catkins, rendering an apparently heavy crop valueless due to poor pollination.

Insect attacks

Insects may sometimes affect seeds of certain broadleaved trees and conifers. Seeds of *Pseudotsuga menziesii* and *Abies* spp. can be attacked in the cone by seed wasps (*Megastigmus* spp.), and *Quercus* spp. may be attacked by *Tortrix viridiana* caterpillars which cause defoliation just after flowering and the knopper gall wasp (*Andricus quercuscalicis*) which attacks developing acorns. Attacks of the latter vary greatly from year to year and may influence the decision of whether the crop should be collected or not.

Insect attacks are rarely significant in years of heavy seed crops (normally infestation levels are far less then 10%), but in light and moderately light crops infestations can reach disastrous levels. This is because the available seeds in heavy crop-years far exceed those required to sustain the endemic insect population, whereas in light crop-years they may be more or less limiting. Therefore, in light crops of *Pseudotsuga menziesii* and *Abies* spp., infestation levels should be checked before finally deciding whether or not to collect.

Another type of insect attack affects acorns at a later stage. It occurs to some extent each year although the attacks may be cyclical. The insects involved are weevils of the genus *Curculio*. The females lay their eggs on the young developing acorns and the larvae hatch and eat into the developing embryos. Outwardly their presence cannot be detected until the larva has finished feeding when it will eat its way out of the acorn leaving the tell-tale hole. It will pupate in the leaf litter and complete its life cycle the following year. Any larvae emerging after acorns have been collected and stored do not re-invade other uninfected acorns in a lot. The larvae finish their eating shortly after seed harvest and fall to the ground away from the acorns. The number observed on the floor under a stack of acorn sacks can cause quite considerable concern particularly as they resemble closely the vine weevil larvae. It should be stressed that they are in no way related to the latter. Apart from ensuring that the acorn weevil larvae are swept up and disposed of no useful treatment can be given to the acorns at this stage. Provided the larvae do not eat the radicle and plumule tip, the acorn does not actually lose its ability to germinate although the insect frass does act as a focus for fungal contamination.

Limited evidence suggests that placing acorns in a cold store slows down the metabolism of the larvae, delays their migration out of the acorns and prolongs their feeding. This possible adverse effect must be balanced against the known beneficial effect of cold storage helping to maintain acorn viability.

Levels of contamination vary greatly from year to year but levels up to at least 20 per cent are regularly observed. It is claimed that in France warm dry summer weather encourages the spread of the insects. As all species of *Curculio* weevils except one are native to Britain and are extremely difficult to identify, weevil infestation is not in itself regarded as a reason for banning acorn importation.

Infestation levels of any seed pest can be determined by sending a representative sample of cones or seeds to the Official Tree Seed Testing Station, Alice Holt Lodge, Wrecclesham, Farnham, Surrey, where seeds will be extracted and X-rayed. The photograph is used to estimate the proportion of sound seed. Alternatively it is possible to check the infestation locally by splitting open the broadleaved seed or cones. The technique for splitting cones is the same as that previously described and illustrated (Figure 7.1). The number of exposed, full and insect-infested seeds are counted – if more than half the seeds are infested by insects the economics of the collection should be carefully examined.

FURTHER READING

ANON. (1982). *Les semences forestières*. Note Technique No. 48. CEMAGREF, Ministère de l'Agriculture, France.

DOBBS., R.C., EDWARDS, D.G.W., KONISHI, J. and WALLINGER, D.P. (1976). *Guideline to collecting cones of BC conifers*. British Columbia Forest Service/Canadian Forestry Service, Joint Report No. 3.

GORDON, A.G. and ROWE, D.C.F. (1982). *Seed manual for ornamental trees and shrubs*. Forestry Commission Bulletin 59. HMSO, London

MATTHEWS, J.D. (1955). Production of seed by forest trees in Britain. *Forestry Commission Report on Forest Research 1954*, 64–78. HMSO, London.

SEAL, D.T., MATTHEWS, J.D. and WHEELER, R.T. (1965). *Collection of cones from standing trees*. Forestry Commission Forest Record 39 (revised). HMSO, London.

Chapter 8

Cone and Seed Collection and Handling Before Processing

by **A. G. Gordon**

Introduction

Successful collection of cone and seed crops will be achieved only if it has been well planned and the necessary equipment, materials and facilities are available on time. If the species is covered by the Forest Reproductive Material Regulations, 1977, notification of intention to collect should be sent to the appropriate Forestry Authority National Office at least 28 days before the start of collection is planned (see Chapter 5). The National Office will acknowledge the notification and request that further contact thereafter be made with exact details to the local Forestry Authority Office. Full details of the current procedures, are contained in a free booklet entitled *The Forest Reproductive Material Regulations – an explanatory booklet* (1987), available from the Forestry Commission, Grants and Licences Division, 231 Corstorphine Road, Edinburgh EH12 7AT.

Conifers

Timing of collection

The date at which cones reach maturity varies between species and with latitude, altitude and season (see previous chapter, pp.72-74). Thus the dates recommended for collection in Table 7.1 should be treated as guidelines which may need to be adjusted to take account of seasonal and geographical differences. For example, collections in the north of Scotland will normally begin as much as a month later than collections of the same species in the south of England, although for some species e.g. *Picea sitchensis* photoperiodism seems to affect the date of onset of seed release. Thus in 1983, 1984 and 1985, seed release had been completed in north Scotland several weeks before it had been completed in England and Wales.

Seeds of many conifers are physically mature long before they are shed naturally and so it may be possible to collect cones up to a month earlier than indicated in Table 7.1. Although cones may be picked early, the seeds cannot be extracted until the cones are fully ripened and so very careful handling and storage is required from the time of collection until extraction. Early collection may be desirable where the crops are so heavy that all the cones cannot be collected in the normal season or where, from experience, damage from crossbills or squirrels may be expected; it should not be attempted unless proper storage facilities can be provided.

Cones of all *Abies* spp. burst open when fully ripe and should be collected while still slightly green in late August to September. Because these cones tend to heat up very quickly after collection they must be stored under well-ventilated conditions (see Figure 8.1). Cones of *Pseudotsuga menziesii* and *Picea* spp. ripen slightly later (September to October) and should be collected when they have lost more of their green colour. Cones of *Chamaecyparis lawsoniana*, *Thuja plicata* and *Tsuga heterophylla* all mature about the same time in September and October and seed is shed quickly as the conelets change from bright green to yellow. *Pinus* spp. vary considerably in the date of maturation of their cones. Cones of *Pinus contorta* should be collected from late August in the south and coastal parts of Britain to October in high inland forests in the north, and cones of other pines from November to March. Cones of *Larix leptolepis* and *L. x euroiepis* mature in September, about one month earlier than *Larix decidua* cones. *In all cases the exact timing of collection must be decided locally by observation of the crop and season.*

Methods of cone collection

The method used depends upon the terrain, spacing between the trees, species and age and management of the stand. The basic decision to be made is whether collection will be from felled trees or by climbing. Properly trained climbing teams can make cost effective collections and such a method should not be dismissed. However, with the exception of young trees, isolated or edge trees, and *Abies* spp., the normal method of collection in Britain and becoming more widespread in the world, is collection from felled trees.

Collection by climbing or from standing trees

In very young pine and larch orchards and seedling seed stands early collections can often be made entirely from the ground. In later years it is essential to have sufficient suitable collection equipment, otherwise collections will be very costly and inefficient. Fruit picking tripod ladders are suitable on fairly level ground for trees up to 6 m tall. Taller trees have to be tackled by other methods and on fairly flat firm sites and where the tree crowns are well separated, hand-propelled, wheeled, vertically-extending ladders such as the Tallescope can be used on trees up to 11 m tall. On rougher sloping sites trees over 6 m tall have to be tackled by climbing, using specialised safety equipment and perhaps extending ladders. Hydraulic platforms mounted on trailers are very efficient to use but are costly to buy and the sites on which they can be used are limited. However, in well-spaced stands, ridesides and where there is a particularly heavy crop, they may prove economical to rent. In Europe, access to the crown of clean-boled trees is often by 'drain-pipe' ladders tied to the bole.

In stands specially managed for seed production (permanently registered stands) collection may be by trained climbing teams. However, when there is an exceptionally heavy cone crop, before full rotation age is reached, consideration should be given to felling some or all of the trees and collecting from the ground. Collections from unmanaged mature or almost mature stands are normally made by felling seed trees since the value of the seed on a heavily-coning tree of high quality will always exceed the value of the timber.

Collection from *Abies* spp. should be done by climbing; felling will shatter nearly ripe cones.

Because of their small sized cones, collections from *Chamaecyparis lawsoniana*, *Thuja plicata* and *Tsuga heterophylla* are impracticable and uneconomical by climbing and are best done from standing edge trees which have foliage near to the ground. Large rakes can be used to scrape the small cones on to tarpaulins. Sometimes heavily coning branches can be cut and the cones removed later.

Other aids to cone collection have been employed in the past and are still employed abroad. Tree bicycles were used widely on the continent and are now used in Britain only for climbing specimen trees with long, clean boles. They have been found to be too cumbersome for collection teams to use efficiently. However, climbing irons are still widely used in Europe. Tree-shakers have also been used successfully in Europe and America in seed orchards although the success seems to depend upon species and make of equipment. They are not recommended for use in Britain (see Faulkner and Oakley, 1971). In Canada helicopters have been employed for inaccessible sites with dome-shaped cone rakes which are manoeuvred over the crowns. They have also been employed to cut off the top few metres of coning fir trees and to transport them to the ground for someone to strip the cones.

Cone collections from all classes of stands should be adequately supervised. When climbing is involved there are very strict safety rules to be followed. (For full details consult the

Health and Safety at Work Act 1974.)

Collection from felled trees

In the last 25 years in Britain, by far the larger proportion of cones has been collected from felled trees, either as part of a normal harvesting operation, as part of a thinning regime or from selected fellings of heavy coning trees. The exact method of operating depends to a large extent upon the site, age of the crop, machinery used for extraction, contractors commitments, availability of labour, etc.

Ideally collections should be made from an area of fellings that is not to be extracted immediately, so that there are no problems of safety for collectors from movement of heavy equipment. However this is not always possible and it is therefore the responsibility of the forester or other person in charge to ensure the complete safety of every worker and collector. With proper liaison and discussion with contractors it is normally possible to arrange fellings in more than one area and to alternate collectors and chainsaw operators between the two. Collectors should be kept clear at all times of areas where heavy equipment is working.

The arrangements for actual collection depend on local circumstances. While reliable casual labour can collect cost effectively, this is not always available.

In an effort to improve the efficiency of collection, the Work Study Branch of the Forestry Commission in 1983 undertook a major study of seed collection methods (Cliffe, 1983).

Collection procedures

During active collection cones are normally placed in a collection bag slung in a convenient position for the hand to reach, around the waist or torso of the collector. These are then emptied as necessary or when changing position. Rigid plastic containers capable of holding a half hectolitre (50 litres) are useful for collections from the ground and for measuring the quantity of cones collected.

The effort put into cone and seed collection will be largely wasted unless cones and seeds are handled and labelled properly from the moment they are removed from the tree until they arrive in the seed extraction and processing plant. If possible cones should be surface-dry when collected and packed. If surface moisture is present they should be given as much ventilation as possible. If cones have had to be collected before being fully ripened, for example green cones of *Picea sitchensis,* they too should not be put into sacks immediately. As with wet mature cones they should be dried further, either by spreading them thinly on a hard surface in a well-ventilated building or open shed with overhead shelter, or by blowing dry, cool or very slightly warm air through a pile of them. Such cones should be turned at regular intervals until dry. Green cones should only be bagged after most of the cones have ripened fully and have become straw coloured. If this is not done, the bagged cones will sweat and go mouldy even when the sacks are stored in well-ventilated conditions.

In general hessian sacks have been used almost exclusively in Great Britain in the past. These have a tendency to rot whenever they are stored even slightly damp. Also the resin on conifer cones has sometimes been found to act like acid and to burn holes in the hessian making them unsuitable for re-use. This problem is overcome in some countries by the use of strong loosely-woven polythene bags. The loose mesh allows the cones to breath and the plastic does not burn. Such sacks can be re-used several times. Solid plastic sacks (fertiliser or similar) should never be used for any but very temporary storage and should certainly never be tied unless they have had a lot of large holes punched in them. Such sacks cause sweating and a rapid build-up of heat and thus loss of quality in the cones.

Storage of cone sacks

Sacks of cones should be stored properly at all times. If cones are allowed to heat up or go mouldy seed extraction may be difficult and the seed quality impaired. Sacks of freshly collected cones should not be stored in bulk for more than a few hours directly on the ground or on a concrete floor. They should be stood upright and separated from each other to allow some air circulation; alternatively they can be stacked two or three deep on top of each other on wooden pallets. These principles apply to all species of conifers but because of their woody nature, cones of *Pinus* spp. can withstand more adverse conditions than cones of, for example, *Pseudotsuga menziesii, Picea* spp. and *Abies* spp. which have higher natural moisture levels.

Sacks of cones awaiting despatch should be kept under cover in open-sided sheds or under tarpaulins rigged between trees to allow free air movement. They should be raised off the soil; this can be done by laying them one deep across two poles suspended by blocks or saw-horses or by standing them on polythene sheets. If the only storage available is in a permanent building the doors and windows should be kept open to permit free circulation of air. A suitable inexpensive store for sacks of cones is shown in Figure 8.1.

Despatch of cone sacks

Cones should be despatched to the extraction plant as soon as feasible after collection using the most rapid form of transport possible. In transit the above principles of storage apply. In particular, green, damp or mouldy cones should not be covered with a tarpaulin unless it is actually raining.

Broadleaves

Timing of collection

The quality of the season plays a significant role in determining the time of seed ripening. Warm, dry summers will hasten ripening by up to one month; cool, damp seasons will delay it

Figure 8.1 Quickly constructed temporary storage shed for cones.

by the same amount. However, the heaviest shedding of *Fagus sylvatica* and, particularly, *Quercus* spp. seeds is after a sharp air-frost. In seasons without early autumn frosts there may be a long slow shedding of acorns which can, despite abundant fruit, even result in a harvest failure due to predation. Seeds of *F. sylvatica* are normally shed in two phases; empty and insect damaged seeds fall first with the mature full seeds falling approximately one week later. In young seed orchards of *F. sylvatica* it is possible to pick the cupules by hand when they have just begun to split and the seeds have turned brown. *Betula pendula* seeds ripen over a very long period from August to late autumn. They are ready for collection just as the catkins begin to turn brown from green but can be collected at any stage until the brown catkins are finally dispersed by wind and rain.

Methods of collection

Seeds of *F. sylvatica* and *Quercus* spp. are normally collected from the ground after they have fallen and so the time spent in preparing the site is more than adequately repaid with time gained in collection. Only in young beech orchards can cupules be economically collected from ladders. In mature stands of these species where the density is 300–400 stems per hectare the ground in autumn is often bare and little preparation is necessary. In more open stands any shrubs and grass should be cleared during summer with weedkillers or cut back to ground level before seed fall. This is absolutely essential if costly tree shaking equipment is to be rented to aid seed fall. Here large lightweight sheets must be spread beneath the crowns of trees to be shaken. The exact method employed depends upon individual circumstances.

On the continent, the use of close mesh plastic nets for collecting beech seed is quite widespread. Very large areas (up to and even over 5 hectares) of *Fagus sylvatica* forests are often covered at a time by one nursery or seed firm for this purpose. The ground has first to be cleared of bushes, etc., which will prevent the nets from reaching the forest floor. This method results in the seed, leaves and a few twigs being collected and prevents cupules and stones, which are difficult to separate from full seed in the cleaning process, from being collected. A further advantage of using nets is that the seeds are held above the soil surface, which prevents harmful organisms infecting the seeds. In France in the mid-1970s there were two good mast years, 1974 and 1976, which were characterised by a high level of seedling mortality caused by the soil-born fungus *Rhizoctonia solani* (Perrin, 1979). Reports of such problems are non-existent from the continent since the widespread use of nets. In Britain, beech seed collected in 1989 from a forest floor in the Lake District, which had received no prior preparation, was found to be infected by *Pythium* species which has caused significant mortality when the radicles reached 1 to 2 cm in length.

The preferred way of collecting acorns for sale is to pick up individual, intact and apparently healthy seeds. This avoids the necessity for large-scale cleaning operations at a later time. In eastern European countries, where labour is cheap, a similar method is used for beech mast. However, in western Europe large-scale collections of seed for sale are normally made by sweeping up the leaves, seeds and cupules from the forest floor and roads and separating out the good seeds later. On flat ground urban road sweepers have even been used to collect debris from the forest floor as quickly as possible; on the continent mobile vacuum seed collectors have also been used successfully to make quite pure collections of acorns and beech nuts.

Another method employed where a relatively small number of individual trees have a crop, is to erect tarpaulins, hessian or similar material, under the trees after first removing the undergrowth. This allows seeds to be swept quickly but will also collect other foreign matter. Tarpaulins however are costly and are not practical for large areas. Provided hessian or fruit netting does not have to be bought specially, these are probably best because they do not hold water.

Seeds of *Betula* spp. can be collected by

stripping the ripe catkins by hand from the branches. This is best done by stripping all catkins and leaves together by running a gloved hand down the length of the twigs. After drying, the catkins split and the leaves can then be separated quite quickly. The seed can also be collected by pruning heavily fruiting branches, or felling trees, and leaving leaves and catkins to dry out before beating them on a clean hessian sheet or tarpaulin to release the seeds. Another method is to suspend a fine net or hessian sheet under a tree in such a way as to collect what falls into the centre, then to beat the branches of the tree with sticks to encourage the seed to fall. A net can be left in position for several weeks. Beating of the tree should be carried out at regular intervals, particularly during dry periods.

Broadleaved seed handling

Unless acorns and beech mast are to be cleaned further the best method of maintaining quality is to sow immediately after collection in protected seed beds. However, if seed has to be handled further the principles already described for handling cones in sacks also apply to acorn and beech mast in sacks. Processing should be started as soon as possible after collection but, while awaiting despatch to the processing plant, seeds should be stored in ways which will allow the seed to breathe without allowing them to lose too much moisture or to heat up. This is best achieved by using wide mesh sacks and standing them individually with ventilation on all sides. They should not be heaped in a pile.

If acorns have begun to sprout by the time they have been collected, which will often happen during a very rainy harvest period, they should be handled as carefully as possible. Sprouting acorns respire more than intact acorns which causes moisture to condense on the outside of sacks and on any covering; this in turn promotes further germination. They should either be given very good ventilation by being put in loose-weave sacks (stacked singly until despatched), or the acorns should be laid out in thin layers on a dry surface and turned (see next Chapter). Alternatively, they should

be bagged and shipped immediately to the processing plant, taking all possible precautions to maintain quality (see below).

Seed of *Betula* spp. collected while the catkins are still slightly green will be at an unacceptably high moisture content immediately after harvest and must soon be dried down to storage moisture levels if it is not to sweat and deteriorate. This is best done by spreading it thinly on a dry surface and blowing air very gently over and, if possible, through it. Until the moisture content has been shown to have fallen to safe levels for storage the bulk seed should be stored in hessian sacks which should be allowed to breathe at all times.

During transport sacks of unprocessed seeds of all species should be allowed to breathe but should not be subjected to drying air currents. They should not be covered tightly with tarpaulin covers for long periods which will cause them to sweat. They should therefore be transported as loosely as possible in a covered vehicle or else should be stacked in an open vehicle and covered by a hessian cover or other permeable covering. They should only be covered by a tarpaulin when it is actually raining.

REFERENCES

CLIFFE, P. (1983). *Study of seed collection methods and output guidance from felled trees.* Southern Region Work Study Report 167. Forestry Commission, Edinburgh.

FAULKNER, R. and OAKLEY, J.S. (1971). *Trials with a mechanical tree-shaker for harvesting cones and beech nuts.* Forestry Commission Research and Development Paper 79. Forestry Commission, Edinburgh.

PERRIN, R. (1979). La pourriture des faines causée par *Rhizoctonia solani* Kühn: incidence de cette maladie après les fainées de 1974 et 1976. Traitement curatif des faines en vue de la conservation. *European Journal of Forest Pathology* **9,** 89–103.

Chapter 9
The Processing of Cones and Seeds

*by **A. G. Gordon***

Introduction

On arrival at the processing plant, sacks of cones and seeds should be off-loaded as soon as possible and stacked in a well ventilated place to allow any heat and sweating to dissipate. In some plants which use a box pallet system of handling cones and seeds, this can be done by emptying the sacks immediately into the pallets but in most cases it is necessary to store the sacks for a period before further processing can take place. This storage is best done in racking made of thin poles which will allow air to circulate around the sacks from all sides. Where squirrels are a nuisance adequate protection must be given to the sacks.

Conifers

There are many different systems used throughout the world for processing cones but all employ heated or dehumidified air in one form or another to reduce moisture content of the cones to a level at which the seeds are released. All systems have shown themselves to be capable of producing seeds of an acceptable quality but little direct evidence is available about the relative efficiency of each. However, on theoretical grounds it has been shown that those where the air passes through or over all of the material to be dried, are more efficient in use of energy than systems where only part of the air current passes through the load. A diagram of the stages in cone and seed processing is provided as Figure 9.1.

European kilns

The kilns used nowadays in continental Europe almost invariably resemble those used over many centuries for cereal and malt kilns. In these, a source of heat is located at ground level or below and hot air rises or is forced upwards through the cones. There are many variations on the basic design but all employ a vertical movement of air.

The most simple kilns employ one single drying floor, which is perforated to allow movement of air but which can be broken into sections to allow the open cones to fall into the seed separator below. In single floor kilns it is also normal to find a means of recirculating the air, if desired. This is done when the heated air has not been fully saturated with moisture and conserves energy significantly. The air that actually carries out the drying may be air mixed with the products of combustion of the fuel or, more normally, it is heated indirectly to reduce the risk of ignition of the resinous dust found throughout such facilities. Kilns of this type are found in the older-established private seed firms in Europe, and have shown themselves to be very effective and to produce top quality seed (see Figure 9.2).

Most recently-constructed kilns have been designed with a series of drying floors arranged vertically above each other and with a cone storage area on top of that. On arrival, the cones are hoisted to the top storage floor, which is well ventilated to allow pre-drying of the cones, and thereafter the movement of the cones from floor to floor is by gravity. In this

Flow diagram of steps in cone and seed processing

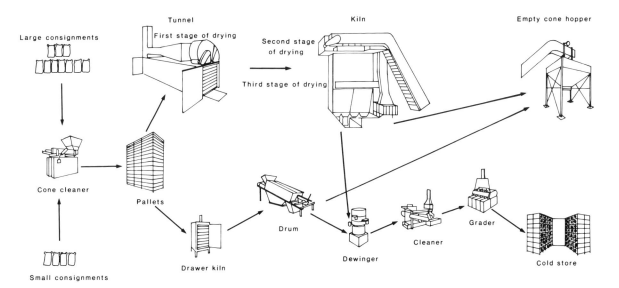

Large consignments

Cone cleaner

Small consignments

Pallets

Tunnel
First stage of drying

Drawer kiln

Second stage of drying

Third stage of drying

Drum

Kiln

Dewinger

Cleaner

Grader

Empty cone hopper

Cold store

Figure 9.1

type of kiln no recirculation of the hot air takes place. It passes up from the furnace or heater through the first perforated floor and on up through each higher floor, gradually becoming more saturated and cooler. It is exhausted from the top floor without passing near the stored cones. As the cones on the lowest level reach the desired condition, they are released into the seed separator by opening sections of the kiln floor. The cones in the next higher level are then dropped to the lower level after it has been re-closed and the process is repeated for as many drying floors as there are. The usual number of such floors is two or three but in one of the largest kilns, built in Czechoslovakia in the late 1960s, four such drying floors were built. In this latter kiln the heat is provided by furnaces capable of running on cones, coal or oil.

In Europe, storage of cones for a period prior to kilning is common practice. This causes a reduction in the moisture content of the cones thus reducing drying costs, but is said also to act as a maturation period for the cones. This maturation period is sometimes called 'after-ripening', but it should not be confused with the same term used to describe the elimination of dormancy in fully processed seeds (see Chapter 11). In most European kilns, cones are stored in bulk in very well ventilated areas, but shallow sided pallets with fine mesh bottoms are also used quite widely. Where pallets are used for storage the drying takes place by placing them in stacks over vents in the floor of the drying chamber. These stacks of pallets fit the chamber exactly, which forces the air from the vents to pass up through the cones.

North American kilns

In North America, the type of kiln normally used is one based upon wood kilns. In these no attempt is made, due to the nature of wood, to pass heated air through the material to be dried; instead the air is circulated around the stack of wood piled loosely in a chamber. Kilns used for drying cones in North America are therefore normally designed in a horizontal axis. Damp cones are placed in pallets and these are put in at one end of the chamber, and

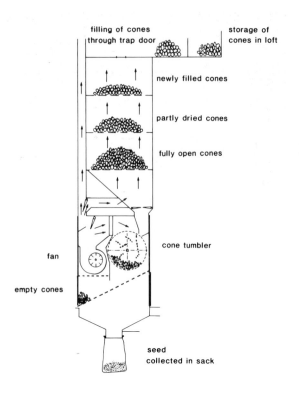

filling of cones
through trap door

storage of
cones in loft

newly filled cones

partly dried cones

fully open cones

cone tumbler

fan

empty cones

seed
collected in sack

Figure 9.2 Schematic representation of a typical
European kiln.

slowly moved through the chamber as the
pallets of dry cones are removed. Heated air
enters the chamber under pressure at the dry
end and leaves the chamber at a lower
temperature and with increased humidity at
the other end. In most such kilns air that has
not been fully saturated can be re-circulated in
order to conserve energy.

Other North American kilns are designed on
the batch principle. Cones are placed in pallets
and these are put into the kiln and removed
when the cones are ready for seed extraction.
There is no systematic progression through the
kiln. Air is recirculated automatically when the
relative humidity falls below a chosen level.

In North America the distances sacks of
cones are transported after collection are often
very great. For this reason they are stored at
the point of collection often for quite long
periods and only transported in bulk when the
collections are finished in order to keep

transportation costs down. Such storage is in
sacks and is often in quite humid conditions.
On arrival at the kiln the sacks of cones are
stacked on racking in large sheds with half-
timbered sides. They are placed in such a way
that air is able to get at all sides. They stay in
the sheds often for some weeks, during which
time the cones begin to dry out (although on
the Pacific coast this may not proceed very far).
In North America this period of 'after-ripening'
is regarded as vital to the production of good
quality seed especially for *Abies* spp. It is said
to allow the embryo in the seed to develop fully.
Coming after storage and transportation over
long distances the logic behind this claim is not
very clear. In view of the generally poorer
germination quality of imported *Picea
sitchensis* seed, compared with that produced in
Britain (see Table 11.4), the claim must be
questioned (Gordon, 1978).

In one Canadian extraction plant a special
cone store has been constructed which
circulates air at 5°C around cones of *Abies* spp.
kept in loosely woven sacks held on modular
racks for several months until the January
following harvest.

The Forestry Commission kiln

The kiln constructed in the grounds of Alice
Holt Lodge, Farnham, Surrey, in the early
1960s incorporated elements of both the
European and American systems. It is a two-
stage process; the first resembles a North
American horizontal axis progressive kiln and
the second incorporates the kiln floor of
European kilns. The two stages are illustrated
in Figures 9.3 and 9.4. The source of hot air is a
double conversion through steam from an oil
burner.

At Alice Holt cones are emptied into pallets
with shallow sides and fine mesh bottoms as
soon as possible after they arrive. These are
stored in stacks of 16, inside the large building
which houses the kiln and processing
equipment. The heat given off from the kilns
results in considerable pre-drying of the cones
in the pallets. Up to 50 per cent of the cone
moisture can be lost depending upon the length

of time stored, which is itself dependent upon such things as date of arrival and the desire to extract all cones of one species together to reduce time spent in cleaning the equipment.

Experiments at Alice Holt Research Station have failed to detect any differences in quality of seed of several species extracted immediately after collection from the quality obtained from those extracted after several months of pre-drying. However, a comprehensive series of experiments did, surprisingly, show a clonal difference in *Pinus sylvestris* in the ease of extraction (judged by the percentage recovery of full seeds in cones using a standard extraction procedure) between cones extracted immediately after harvest in November and cones from the same clones extracted immediately after harvest in March. Clones which released almost all their seed easily in November did not release so much in March and *vice versa* (see Buszewicz and Gordon, 1973; 1974; 1975; Gordon, 1976).

At Alice Holt stacks of eight pallets are placed in the cooler end of the chamber at 30°C. As stacks with dry cones are removed from the warmer end those in the chamber are wheeled forward, becoming exposed to progressively warmer and drier air. Timing of removal of the cones is determined by experience of the opening and is not for any set time. The warm and almost fully-open cones are then tipped into a conveyor which transports the cones and any extracted seeds to the box kiln on the top of the second stage (see Figure 9.4). Here air at a warmer temperature (40°C) is forced in from the side and up through the finely perforated base of the box and then exhausted. After a period of further drying, which is mainly determined by the time for the next stage to be completed, the open cones are dropped through the open trap door of a revolving drum made of slatted metal sheet, the slats being big enough to allow all seed sizes to pass through, but also small enough to retain the empty cones. The drum is revolved slowly until no more seeds are seen to fall from the cones. The chamber in which the drum revolves has even warmer air (50°C) passing through it. It is this air which then passes through the floor of the box kiln.

Figure 9.3 The tunnel stage of the Alice Holt kiln. (*B9392*).

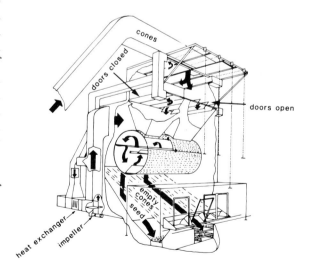

Figure 9.4 Diagrammatic representation of tower stage of Alice Holt kiln.

All seeds extracted in this way are collected in the bags at the bottom, while spent cones are exhausted from the drum by opening the traps, and then by conveyor to the outside of the building.

The EFG kiln

Much research and development on drying of agricultural crops has taken place at the National Institute of Agricultural Engineering at Silsoe, since the Alice Holt kiln was installed. Many different types of kiln have been designed but all have used the principle of moving heated air through the material to be dried. When the Nurseries Division of Economic Forestry Group plc wanted to build a small and adaptable kiln in 1983 they were advised by the NIAE that the most appropriate design was one based upon a drying wall. A simple kiln complete with heater using LPG (liquid propane gas) and revolving drum separator was constructed for approximately £1500. This has worked well at costs per kilogram of extracted and cleaned seed, very comparable with those of much larger units. Yields of seeds per hectolitre of cones and quality have been totally comparable with those of much larger kilns.

The Forestart kiln

In 1987 when building a new kiln, a completely new technology, using the principle of the heat pump as the source of energy was identified as being the most energy efficient. Moisture from the material to be dried is taken up into dry air moved over it by heavy-duty fans, is condensed on a cooling coil in a dehumidifier releasing latent heat of condensation and recycling the dry air. The layout of the kiln is shown in Figure 9.5. For a modest outlay of £5000 a kiln capable of processing 25 hl of cones in a 24-hour period was produced. The efficiency was quite startling with energy saving of 70–80 per cent compared with the energy used by the EFG kiln, while the latter was itself shown to be more efficient than the double-phase conversion kiln in operation at Alice Holt Lodge since the early 1960s. Seed quality from the

Figure 9.5 Forestart kiln using heat pump as source of cheap and efficient energy. (*Photo: John Poole Photography*).

Forestart kiln has been consistently high and the system has the added advantage that drying can be done by dehumidification at temperatures below 30°C which is highly desirable for more sensitive species such as *Abies grandis* and *Abies procera*.

Physical parameters in cone extraction

Any source of heat either direct or indirect can be used for drying cones, but whichever source is used care has to be exercised in the design of the equipment to prevent ignition of the highly combustible conifer dust. Therefore in most kilns indirect heat is used. Cones are a valuable source of energy and are used quite widely in many parts of the world. All other fuels are used but electricity is the least employed because of its high unit cost.

Temperatures used in drying must be related to the condition of the cones. Soon after harvest, when the cones are green and have a moisture content near 50 per cent, the temperatures should not exceed 30°C. Later as the cones dry to about 20 per cent moisture content, the temperatures can be increased to nearer 50°C. Providing the moisture content of both cones and seed is low enough, temperatures even as high as 60°C for a short period will not adversely affect seed quality. However, outside Britain great care is taken that the air temperature does not exceed 45°C. Normally the higher the temperature the faster the extraction *but the temperature should always be built up gradually.* High-temperature extraction is not always the most economical method and only experience will show how best to use the equipment available, and how to determine when the cones have dried and are open enough to release their seed fully. It is not very practicable or necessary to use the moisture content of cones as an indication of cone opening, because there is no absolute relationship between this factor and the reflexion of the scales. Also there appears to be clonal variation in some species, for example *Pinus sylvestris,* in the proportion of scales in a cone that actually reflex. Additionally this property seems to vary from year to year, possibly depending upon the level of pollination.

Separation of seed from cones

In all kilns the final process of extraction involves tumbling the opened cones within a perforated drum, which allows only the seeds to fall through the perforations. Provided the cones are fully open this procedure presents few problems and almost all seeds are extracted. However, some genera exhibit particular problems. Thus most *Larix* spp. cones and in particular lots with large amounts of resin, are very difficult to extract and require prolonged tumbling in a revolving drum, which allows the cones to be ground down against each other while releasing the seeds. Alternatively the seeds may be shaken in a special reciprocating drum or they may be released from the cones by grinding the cones lightly in a hammer mill.

The cones of *Abies* spp. also require special apparatus since they disintegrate on drying and the seeds must be separated from the similarly shaped scales. The perforations in the revolving drum must be big enough to allow the winged seed to pass through while still retaining the seed scales. Because of the similarity in size of the seed and scales, extraction and cleaning of *Abies* seeds can be time-consuming.

Cones of some pine species exhibit case-hardening when kilned. In this the scales remain tightly closed even after drying completely and the seeds are retained. It can be overcome by remoistening the dried cones, soaking them overnight in water, followed by rekilning. Some species, e.g. *Pinus contorta* in central Canada, require special solvents, or heating at high temperature for short periods, either dry or in water, to break the resin binding the cone scales. Such cones are known as serotinous.

The extraction of seed from small consignments can be carried out in any warm situation, but is best done in a forced-air oven. After drying the cones the seeds may be separated by shaking them in coarse sieves or rotating them in special wire-mesh drums.

Seed cleaning

A common characteristic of all conifer seed is the possession of a large or small wing. For ease of storage and sowing of most species this wing must be removed. The method depends upon the anatomy of the seed of the various genera involved. *Abies, Larix* and *Pseudotsuga* seed possess large wings which form an integral part of the seed coverings. *Picea* and *Pinus* seed have wings which are attached to the seed but which are separate and detachable structures. *Chamaecyparis, Thuja* and *Tsuga* seeds have very fine wings which do not need to be removed for economic storage and easy sowing.

For many years up to the late 1960s dewinging of *Picea* and *Pinus* seeds was carried out using brush dewingers. In these, winged seed passed between one static and one revolving set of brushes. The machines were not very efficient; they either left a lot of wings on the seeds or damaged the seeds if the dewinging was complete. *Abies, Larix* and *Pseudotsuga* seeds were dewinged by continuous movement of the seeds against each other in a machine resembling an agricultural auger set vertically. By bringing the seed to the top and allowing it to fall again with gravity while moving against each other, the wings were gradually abraded over a period of hours.

In the 1960s, on both sides of the Atlantic, workers discovered that *Picea* and *Pinus* seeds could be very efficiently dewinged by wetting the wings slightly. This produced a natural release-reaction in the cup or caliper holding the wing to the seed and resulted in extremely well dewinged seeds. In America and in some European countries the wetting has been carried out in an unmodified cement mixer. About one volume of water is added to four or five volumes of seed and the whole is mixed for about half an hour. In this time the great majority of seed is dewinged. At Alice Holt, a rather more sophisticated machine was constructed which produced equally good quality seed. A vibro screen, normally used for industrial purposes, was equipped with a spray that allows seed to be wetted as it is fed into the vibrator. After a short period of vibration the wet wings fall away from the seeds. The Alice Holt dewinger is also used for dewinging seeds of *Abies, Larix* and *Pseudotsuga*. For these species it is used dry and the vibration of the seeds against each other gradually abrades the wings. However, in the absence of any other purpose-built machine a cement mixer has proved to be a satisfactory tool for dewinging seeds of the same three genera in the dry state.

After wet dewinging the seeds must be re-dried as soon as possible. Any system allowing slightly heated air to pass through and over the seeds is satisfactory. As the seed is only surface wet, it dries down to its original moisture level very quickly. At Alice Holt, this process is accelerated by 'fluidised' drying by using a warm air blower unit coupled to the dewinger (see Figure 9.6).

Figure 9.6 Vibrating separator used at Alice Holt for dewinging seeds. (*B9394*).

Figure 9.7 Pre-cleaner used at Alice Holt. (*B9393*).

Table 9.1 Cone, seed and yield data of commercial forest trees.
Compiled from data gathered from Forestry Commission extraction plant, Alice Holt Lodge, and Official Seed Testing Laboratory, Alice Holt Lodge between the years 1970 and 1986, and Technical Note No. 48, *Les semences forestières*, CEMAGREF, Ministry of Agriculture, France (full address on p. 36).

Part I. Conifers

Characteristic		Abies alba	Abies grandis	Abies procera	Chamaecyparis lawsoniana	Larix decidua
Weight of one hl of fresh cones		45-55 kg	45-55 kg	50-60 kg	35-45 kg	25-35 kg
Average number of cones/hl		400	700	220	107,000	9900
Yield of seed from one hl of fresh cones (g/hl)	lowest		325	695	3590	310
	average	4000	1460	1550	3930	550
	highest		1790	1830	4635	720
Average number of pure seeds per kg		22,500	45,000	40,000	460,000	170,000
Average germination percentage		45%	40%	35%	50%	40%
Average number of germinable seeds per kg		4000	20,000	15,000	230,000	70,000
Approximate number of plants produced per kg of seeds		3000-5000	4000-6000	5000-8000	60,000-80,000	15,000-20,000
Normal successful storage period for seeds above 0°C		2-3 years	3-5 years	3-5 years	3-6 years	3-6 years

Characteristic		Larix leptolepis	Larix x eurolepis	Picea abies	Picea sitchensis	Pinus contorta
Weight of one hl of fresh cones		23-33 kg	25-35 kg	50-60 kg	30-40 kg	40-50 kg
Average number of cones/hl		9600	8500	960	3600	4400
Yield of seed from one hl of fresh cones (g/hl)	lowest	490	330	875	600	190
	average	900	945	1250	930	690
	highest	1310	1540	1510	1255	1270
Average number of pure seeds per kg		250,000	210,000	145,000	400,000	300,000
Average germination percentage					90%*	
		40%	30%	80%	60%**	90%
Average number of germinable seeds per kg					360,000*	
		100,000	60,000	110,000	240,000**	270,000
Approximate number of plants produced per kg of seeds		25,000-40,000	15,000-20,000	35,000-45,000	100,000-150,000	90,000-115,000
Normal successful storage period for seeds above 0°C		3-6 years	3-6 years	8-10 years	8-10 years	8-10 years

* British seed
** Imported seed

Conifers *continued*

Characteristic		*Pinus nigra* var. maritima	*Pinus sylvestris*	*Pseudotsuga menziesii*	*Thuja plicata*	*Tsuga heterophylla*
Weight of one hl of fresh cones		40-60 kg	40-55 kg	40-55 kg	—	40-55 kg
Average number of cones/hl		2800	5500	3000	195,000	58,000
Yield of seed from one hl of fresh cones (g/hl)	*lowest*	810	260	265	1030	—
	average	1115	530	400	1465	1350
	highest	1380	805	610	2065	—
Average number of pure seeds per kg		70,000	165,000	88,000	850,000	650,000
Average germination percentage		80%	85%	80%	60%	65%
Average number of germinable seeds per kg		55,000	145,000	70,000	500,000	420,000
Approximate number of plants produced per kg of seeds		15,000-25,000	50,000-65,000	25,000-35,000	70,000-100,000	80,000-110,000
Normal successful storage period for seeds above 0°C		8-10 years	8-10 years	5-8 years	2-4 years	2-4 years

Part II. Broadleaves

Characteristic		*Betula pendula*	*Fagus sylvatica*	*Quercus petraea*	*Quercus robur*	*Quercus rubra*
Weight of one litre of fruit		100–200 g	400–500 g	500–700 g	600–800 g	500-700 g
Number of seeds per kg	*lowest*		3000	250	200	200
	average	1,900,000	4600	315	270	280
	highest		5000	400	350	400
Germination percentage after harvesting and cleaning	*lowest*	10	—	70	70	80
	average	26	—	79	81	82
	highest	45	—	90	95	95
Viability percentage after harvesting and cleaning	*lowest*	—	60	—	—	—
	average	—	68	—	—	—
	highest	—	95	—	—	—
Average number of germinable (viable (v)) seed per kg of seeds		150,000	3000 (v)	250	220	240
Approximate number of plants produced per kg of seeds		30,000-50,000	1000-1500	80-120	80-120	80-120
Normal successful storage period for seeds above 0°C		2-3 years	1-6 months	1-2 months	2-3 months	3-6 months

Notes   *Fagus sylvatica:* Successful storage period may be extended by special techniques – see Chapter 10.
     Quercus petraea, Quercus robur, Quercus rubra: Acorn storage is difficult and special facilities are required
     – see Chapter 10.

When dry, the bulk of seeds must be cleaned to remove all debris, needles, cone parts, empty seeds, resin, etc. Any machine incorporating two or more sieves and winnower should be capable of cleaning seeds effectively (Figure 9.7). Many very old and very efficient pieces of equipment are still in use in extraction plants throughout the world. Modern equipment designed for use with agricultural seeds can also be used very effectively. Cleaning should continue until no more debris can be seen easily in the bulk and until a cut test of a sample of seeds shows an acceptably low level of empty seeds. For *Picea, Pinus, Pseudotsuga, Tsuga*, 2–3 per cent of empty seeds is normally acceptable. For *Larix, Chamaecyparis* and *Thuja* seeds, where the density of full and empty seeds is nearly the same, a much higher percentage of empty seeds is acceptable. (If cleaning proceeds too far, too high a number of full seeds will be discarded.) A quick and more accurate method of assessing the number of full and empty seeds is by X-ray as described in Chapter 11. Before putting seed into individual containers the bulk should be thoroughly homogenised using special equipment (see Figure 9.8) or by repeated tipping into a number of suitably sized drums.

Details of the yields of clean seed expected for each species are given in Table 9.1.

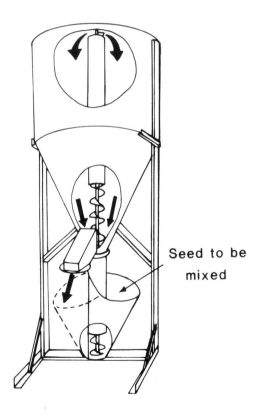

Seed to be mixed

Figure 9.8 Seed mixer.

Broadleaves

Seeds of *Quercus, Fagus* and *Betula* spp. have each to be handled differently.

Quercus

At time of collection seeds of *Quercus* spp. will usually still show some green colour and will have a moisture content in excess of 50 per cent fresh weight. They will be respiring strongly, giving off heat and water as they do so. When placed in sacks and stored for any length of time, the bulk will heat up quite rapidly and the warm moist conditions will begin to cause germination in non-dormant species (*Q. petraea* particularly, and less commonly *Q. robur*). For these reasons the period during which the acorns are kept in the sack before being processed should be kept to an absolute minimum. The object of processing is to slow down the respiration which is mainly centred in the testa, while maintaining the moisture content at the relatively high level of 40–45 per cent.

This is best done by storing the seeds in layers up to 30 cm deep on a wooden floor, in a place where the temperature will not be too high but where the ventilation is good. They should be monitored each day and turned by means of a wooden shovel to keep the drying uniform. Water from sweating should not be allowed to build up as it becomes a focus for fungal attack. When the loss of moisture begins to slow down (this can be gauged by daily weighings of a sample in a loose-weave nylon net) the seed will have reached a moisture content of about 40–45 per cent. The respiration will have tailed off and any further significant moisture loss will be at the expense of the embryo and will be potentially dangerous. When the acorns have lost all their green colour, the seed coats are hard to the finger nail and the embryos still completely fill the cavity. They are then ready for bagging. If this cannot be completed immediately the acorns can be stored on the wooden floor but the depth of the pile can be increased to 40 cm in an effort to reduce further moisture loss. During the turning and bagging processes it is normally possible to sieve the seeds to remove leaves, twigs and other foreign items. This will also ensure that the seeds are thoroughly mixed, which is often important in these large seeded species where there is the danger that seeds from only one tree will find their way into individual bags.

For smaller collections the same principles apply, but it is normally possible to effect the drying and ripening of the seed within the sack, provided it is loosely woven and kept in a cool well-ventilated place. At some time in the drying it will be necessary to mix acorns in order to ensure uniformity.

In wet autumns, acorns of particularly *Quercus petraea,* begin to germinate very soon after they have fallen to the ground. They should be sent to the processing plant as soon as possible (see p.85) where they should be specially handled in order to minimise the amount of sprouting that takes place. Even during dry storage acorns of *Q. petraea* have a greater tendency to sprout that those of *Q. robur.* Unless they are stopped from growing they will continue to respire profusely and in bulk this will cause a further build up of moisture which in turn will lead to more germination and so on. Every effort should therefore be made to stop the continued growth of the sprouted roots. For very large quantities of acorns this requires enormous space and this is seldom available. In such circumstances the acorns are best sent as quickly as possible to the user nursery where they can either be sown immediately or else, because of their smaller bulk, more easily treated.

Sprouted acorns should be spread as thinly as possible in order to dry them out as quickly as possible. Dry air (and even slightly heated air can be used) should be blown over them with the acorns being regularly turned. This should continue until the sprouted root has gone brown and dried up completely. This will not kill the acorn as *Quercus* spp. have the ability of putting out secondary roots which completely replace the function of the primary roots. In fact plants grown from such acorns do not possess the strong tap roots of normal *Quercus* plants which is sometimes regarded as a benefit. Unless badly sprouted acorns have

their roots killed back in this way before sowing the resulting plants will have very distorted roots making them frequently unsaleable.

Greater care is required when drying acorns of *Q. petraea*. Over-drying of the radicle often leads to infection spreading up the embryo to the plumule, resulting in seedling failure. For this species the tip of the radicle alone should be dried with more frequent turning than for *Q. robur*.

After the roots have been dried out the drying process should be terminated in case the acorns themselves begin to lose excess moisture and so become at risk from loss of viability. At this stage they can be bagged up and treated as normal acorns.

For statutory seed testing purposes a bulk of 5000 kg of *Quercus* seed can be treated as one unit. However, it is very difficult to gauge if a bulk of this size is completely homogeneous, a condition that is essential before sampling for seed testing purposes can take place. Thus it is better to sample the bulk after it has been bagged rather than before. For this the procedure described in Chapter 11 should be used.

Fagus

The moisture content of *Fagus* seeds at time of harvest will be quite high (about 40 per cent) and therefore if stored in bulk they will respire heavily and begin to heat up. The object of processing therefore is to reduce the moisture content to an acceptable level (about 25 per cent) and so reduce the respiration and where necessary to clean the seed. However, the amount of cleaning necessary will depend upon the ultimate use of the seed.

Thus seed for immediate autumn sowing may not require much cleaning. Some cupules, twigs, etc., can be left in the seed and can be sown together with the nuts in protected seedbeds. On the other hand seed to be stored until spring or for long-term storage must be cleaned to reduce the danger of contamination from fungi in the debris.

It is not easy to separate nuts from the cupules, twigs and other debris by sieving and for this reason hand collection of the pure nuts is sometimes undertaken. By using a collection of sieves of various sizes and a winnowing machine, a fairly pure bulk can be produced. Modern designs of specific gravity tables have been found to separate full and empty seeds from stones, twigs and leaves very efficiently. Flotation techniques can also be used successfully for processing *Fagus* seeds but they have the disadvantage that they produce a large bulk of very wet seeds that have to be re-dried. A 100 litre tank with a large surface area is used and this is filled with water. Into it a mixture of seeds, cupules and debris (about 10 litres at a time) is placed. The full seed, stones and soil will sink to the bottom whereas the empty damaged seeds, twigs, leaves and expanded cupules will float. Using a sieve these floating objects can be removed easily. As the bulk is processed in this way, so the tank will fill up with seed and other heavier matter. The tank will thus have to be emptied of water and the seed and heavy material removed. The same procedure is repeated until the whole bulk has been processed. The half clean seed can then be sieved in water to remove small stones and soil; then they must be dried. Because of the great difficulty of removing stones of the same size as beech nuts, the time spent in ground preparation to remove these stones is very well spent. (This is one of the great advantages of collecting using nets or hessian sheets.) Drying is best done by laying the seed on a large flat surface and blowing air over and, if possible, through the seeds. The bulk should be turned repeatedly. This is continued until all the surface moisture has evaporated. Cleaning can then be completed by using a normal seed winnowing machine. Final drying can then take place if it is decided that this should be done, but this will depend upon whether the seed is to be stored or whether it is to be sown immediately or treated for sowing in the spring.

Betula

By the time *Betula* seed has reached the processing plant it should already have been dried down to safe storage moisture. This should be done in the way already described (see p.85). At this time any catkins which have

remained intact can easily be broken apart by rubbing with the hand or with a brush. When sufficiently dry (this can be judged by check weighing a sample held in a fine mesh sack until constant weight is achieved) the seed can be sieved to remove any debris, after which it is ready for storage.

It is possible, by very careful choice of sieves and manipulation of air currents in a densimetric cleaning column, to clean seed of *Betula* spp. still further. The bracts of the catkins have a slightly different density from the seeds which therefore can be separated out. This is sometimes done by American seed houses but not generally in Europe.

REFERENCES

BUSZEWICZ, G.M. and GORDON, A.G. (1973). Scots pine seed extraction. *Forestry Commission Report on Forest Research 1973*, 29. HMSO, London.

BUSZEWICZ, G.M. and GORDON, A.G. (1974). Scots pine seed orchard trial. *Forestry Commission Report on Forest Research 1974*, 8. HMSO, London.

BUSZEWICZ, G.M. and GORDON, A.G. (1975). Scots pine seed orchard extraction and germination trial. *Forestry Commission Report on Forest Research 1975*, 7. HMSO, London.

GORDON, A.G. (1976). Scots pine seed orchard extraction and germination trial. *Forestry Commission Report on Forest Research 1976*, 7. HMSO, London.

GORDON, A.G. (1978). The effect of seed maturity on the dormancy and the viability of British produced Douglas fir and Sitka spruce seeds. *Proceedings of the Symposium on flowering and seed development in trees*, 362. Mississippi State University.

Chapter 10
Seed Storage

by **A. G. Gordon**

Introduction

The storage of seeds of commercial forest species has received considerable attention over the years. In the last two decades a very significant amount of research has been done on storage of economically important food and raw material-producing trees, such as rubber. In all this work seeds have been found to fall into two main groups, so-called orthodox and recalcitrant species (Roberts, 1973). Orthodox seeds are those which can be stored successfully for long periods at low moisture contents and temperatures. Within certain limits the lower the temperature and the lower the moisture content, the longer the period of viability. Recalcitrant seeds are those which do not follow these rules, and which are killed if their moisture content is reduced below some relatively high value. For some (mainly tropical tree species), low temperatures *per se* are also detrimental to viability.

From all the work done on tree seeds by the Forestry Commission Research Division's Plant Production (formerly Seed) Branch and elsewhere, nothing has yet been found to suggest that they do not also fit into this scheme. Reports of anomalies referred to in text books have often been found to be unsubstantiated when examined critically, e.g. *Populus* and *Salix* spp. (Zasada and Densmore, 1980). Many have been shown to be caused by poor handling of the seed between tree and the start of the storage treatment.

The Forestry Commission's Plant Production Branch has many years of experience of storage of seed of *Quercus*, *Fagus* and *Betula* spp. and of many commercial coniferous species. Of the species covered in this manual only *Quercus* spp. have been found to be recalcitrant although *Fagus sylvatica* requires special care when being dried down for long-term storage.

Many books have been written on the subject of seed storage. For this reason no further details of the principles will be given here. Readers wishing further amplification are referred to the literature cited at the end of the chapter. Details of the classification of genera of the seed of forest conifers and broadleaves are given in Appendix 10.1.

Orthodox seeds

Most conifers

For long-term storage, seeds of all coniferous species covered in this Bulletin should be dried down to 5–10 per cent moisture content and held between +5°C and −18°C. The nearer to 5 per cent the moisture content is lowered, and the nearer to −18°C the seed is held the longer the viability will be maintained. For practical storage for up to five years it is difficult in most cases to justify using a temperature much lower than 3–5°C, although for some *Abies* spp. with seeds of high oil content, a lower temperature might be justified. For normal storage therefore the temperature of a transplant cold store which is normally held at between +3 and 0°C is ideal. To maintain the moisture content at the desired level, the seeds should be placed in sealed containers. These can be glass or plastic bottles or plastic bags tied tightly by twisting the necks and securing

Figure 10.1 Inside Alice Holt cold store showing racking for efficient storage of seed. (*B9329*).

with wire. Plastic of 400–500 gauge (10–13 μm) should be used. Thinner material allows appreciable exchange of moisture, thicker material inhibits gas exchange (Bonner, 1978).

Although satisfactory storage can be achieved as described above, large-scale storage of many species is best done by putting the closed plastic bags in airtight metal cans. These can then be stored economically of space in the way carried out at the Forestry Commission Seed Store at Alice Holt Lodge as illustrated in Figure 10.1.

Using this system of storage, many species of conifer in the Alice Holt Lodge cold store have shown no significant loss of viability over a period of storage up to 10 years. For some experimental seed lots a successful storage without significant loss of viability of over 20 years has been obtained. For storage data of conifers see Appendix 10.1.

Betula pendula

Results obtained over many years at Alice Holt Lodge have shown that in all respects seeds of *Betula pendula* and all other *Betula* spp. behave in the same way as those of orthodox conifer seeds described above.

Problem species
Fagus sylvatica

Although seed of *Fagus sylvatica* has been shown quite clearly to behave in an orthodox manner, the relationship has been difficult to confirm because of the interaction between the moisture content of the seed, dormancy of the seeds, the loss of dormancy by moist, cool storage and the difficulties experienced in laboratory assessments of the germination and viability percentages. The different storage methods used for beechnuts in different countries of Europe have further complicated the interpretation.

The main reason for the difficulties appears, from experiments at Alice Holt Lodge, to lie in the observation that the Tetrazolium test, as currently carried out, does not give an accurate indication of recent seed damage, so that seed lots treated badly will show a relatively high viability for some time after they have actually lost the ability to germinate. However, as this latter characteristic is measured by a test lasting up to 26 weeks in cool, moist conditions, during which the damaged seed deteriorates completely, the exact moment of loss of germinability is very difficult to identify. Nevertheless, French experience shows quite clearly that newly harvested seed handled carefully and dried slowly over 24–36 hours in moving air at 20°C from 50–55 per cent moisture content at harvest, to 20–25 per cent during handling and cleaning and finally down to 8 per cent, can be stored for many years without significant loss of viability. The lower the temperature the better the maintenance of viability but for practical nursery storage this should not be more than 2°C and less than −5°C (Bonnet-Masimbert and Muller, 1975).

For seeds which are to be sown the spring following harvest it is not necessary to dry the seeds to this low moisture content. They can be safely stored at 20–25 per cent moisture content at 3–5°C until time to begin seed pretreatment prior to sowing. This can be done in a number of ways. Seeds can be stored in sacks in a transplant cold store at between +5°C and 0°C or any shed which remains cool throughout the winter and in which the seed can be protected from rodents. In this type of storage it is advisable to check-weigh a known sample at regular intervals from receipt to ensure that the moisture content remains more or less constant so that its dormancy is not increased (see Chapter 12). Seeds can also be stored in wooden boxes in similar cold situations but this will cause the seed to dry out and this must be corrected by sprinkling lightly with water and turning the seeds regularly. Water can also be applied to a sheet of hessian covering the sacks if these show signs of losing excessive moisture. Alternatively seeds at this relatively high moisture content can be maintained in suitable conditions by being laid between layers of coarse, dry sand.

Another method of maintaining the viability of *Fagus* seeds is to sow them with adequate protection in the nursery beds where they are to be grown. However, this can give problems when spring frosts catch seedlings germinating earlier than expected due to early warm weather and when late spring frosts catch seedlings which have germinated at the normal time, although such damage can be avoided by adequate frost protection.

In most years the seed available in Britain is of continental origin. By the time it has reached Britain its quality may already have deteriorated, and will depend to a very large extent on the treatment to which it has been subjected. Even though it may have a test certificate showing a high figure this may be misleading as it may only reflect the quality shortly after harvest when samples were taken for seed testing. Seldom will the certificate give a moisture content. For this reason it is important, particularly if long-term storage is being contemplated, to obtain an accurate measurement of moisture content. During transit, seed of low moisture will probably suffer little, whereas seed at 20–25 per cent moisture content may suffer significantly if relatively high temperatures are experienced. This can be overcome by insisting upon the use of refrigerated transport.

Unless the exact history and current quality of imported seed lots are known, long-term storage is not advised. Home collected seeds should give better results than imported seed provided their quality has been monitored throughout the handling and is satisfactory.

The methods used for long-term storage differ widely in European countries. In Poland and France it is customary to dry seeds down to approximately 10 per cent moisture content immediately after harvest, and to place them in sealed containers for storage at −5 to −15°C. However, in Germany long-term storage is achieved by keeping the seeds at between 20 and 25 per cent moisture content in peat in sealed plastic bags at −4 to −5°C or in hessian sacks at 25–30 per cent at the same temperature.

The Germans recommend a sliding relationship between temperature of storage and moisture content. The lower the moisture content the lower the temperature. All methods have reportedly given good results although almost always not as good as for fresh seed. Almost certainly the success of the techniques depends upon the quality of the seed at the time the storage treatment begins. For successful long-term storage the very best quality seed should be used.

Recalcitrant seeds

Quercus spp.

It is impossible to store seeds of *Quercus* species for long periods without loss of viability. Indeed, only by careful control of the conditions can they be stored without loss of viability until required for sowing in the spring following collection. The best method of storage is thus to sow them with adequate protection against vermin in the nursery beds in which they are to be grown. Freshly harvested seeds of some species may possess dormancy (e.g. *Quercus rubra*) and the natural conditions in the seedbed over winter not only overcome this, but provide the best conditions available for maintaining the viability. However, good maintenance of viability will take place only if the seedbed conditions are satisfactory. For example, areas likely to suffer from waterlogging must be avoided.

Because of the large volumes involved this practice is widely followed in continental countries. However in years where severe winter temperatures are experienced and where no snow cover protects the seedbeds, very considerable losses can be experienced – up to 80 per cent in 1985/86 and 1986/87.

Bonner (1978) has found that the deeper the dormancy of some *Quercus* spp., the more successfully they can be stored for long periods. This may explain why acorns of *Q. rubra* have been reported to store for up to 5 years whereas *Q. robur* and particularly *Q. petraea*, which are often collected with radicles emerging, will not store for more than one year under the same conditions. The best conditions for artificial storage of recalcitrant species are thus as near as possible to those found in the nursery bed in winter without actually allowing germination to proceed (or at most to proceed too far).

For *Quercus* species these conditions can be provided either by digging a pit and covering the acorns with soil while leaving small straw airways for them to obtain a little air, or by mixing the acorns with an almost dry peat medium and placing them in containers which retain the moisture but allow ventilation. This is time and space consuming for anything but small quantities and is not widely practised commercially.

Without very special storage facilities (see below) the only practical method of storing large quantities of recalcitrant seeds is in sacks. These must be made of material loose enough to allow breathing, but must not be so loose that the moisture content falls too fast. In order to monitor the loss of moisture from stored sacks, several, chosen at random from the bulk, should be weighed at regular intervals. If the initial moisture content is known, the gradual change in moisture content can be followed quite accurately. Most moisture will be lost from the outside seeds in the sacks and these will dry out to near critical levels before the overall moisture content falls very much.

To keep the loss of moisture as low as possible, the sacks should be stacked in a cool place where there is little circulation of air. A transplant cold store operated at high humidity is the best possible place. Storage for up to 6 months with germination maintained at above 50 per cent has been achieved in hessian sacks on pallets containing one and a half tonnes. Direct cooled transplant stores where relatively dry air is circulated are not so satisfactory for long-term storage of acorns as they cause loss of moisture. By preventing the moving air from hitting the sacks directly, the loss of moisture can more or less be controlled and such stores have also maintained germination in excess of 50 per cent for 6 months. A shed on the north side of a building can also give satisfactory results. Here sacks should be placed in stacks two wide and three or four deep so that at least some part of each sack is exposed to the atmosphere. If the moisture loss becomes excessive the sacks should be lightly sprayed with water until the weight of the checked sacks rises to the desired level. When storage is in an outside shed care must be taken that on very windy days the loss of moisture is not excessive. It can be reduced effectively by covering the sacks with hessian sheeting and keeping this wet. If at any time during storage there is any evidence that the seeds are heating up, the sacks should be restacked separately or, if excessive, the contents spread out and reprocessed. During any subsequent transportation the sacks should be covered to minimise excessive moisture loss.

In recent years workers in France have developed a short hot water treatment of acorns to prevent the spread of the lethal storage fungus *Ciboria batchiana*. Although costly to perform, this treatment has maintained very high germination levels (in excess of 80 per cent) until sowing in April. The treatment has resulted at times in epicotyl dormancy so should not be employed without careful consideration of the alternatives.

Long-term storage of acorns

The above recommendations concern relatively short-term storage. Much work has been performed in Poland, France and USA to discover methods of storing acorns successfully for a number of years. Comparisons of methods have shown that not all species of acorns behave in the same way; thus acorns from southern USA do not respond as acorns from Europe or northern USA do.

The most complete studies have been undertaken at the Kornik Arboretum in Poland by Suszka and Tylkowski (1980). They have shown that by careful control of the gaseous atmosphere (enhanced levels of carbon dioxide of the order of 7% with reduced levels of oxygen $c.11\%$) acorns can be stored at $-3°C$ over three winters without reducing the percentage of seedling production to below 60 per cent of the seed sown. Over two winters the germination in the nursery did not fall below 80 per cent. In order to maintain the gas mixture at the required level it was found best to store the seed in a solid container with a perforated tube penetrating to the bottom and the lid left loosely covering the opening. A 50 litre plastic rectangular container with a screw-on 15 mm lid has proved ideal as it maximises the use of space and with racking allows a cold store to be more or less completely filled. Although such storage can be carried out successfully it is not cheap to do and should only be undertaken where it is of paramount importance to have seed of the same origin over a number of years.

In Britain, Gosling (1988) has proven by experimentation what some nurserymen have found out by hard practice. Soaking acorns for 48 hours in water at 2°C significantly improved the germination of acorns that have been stored for up to 28 weeks in cold store at 2°C at reduced moisture levels of 25 to 45 per cent. He also showed that soaking acorns in water prior to storing them in loosely-tied plastic bags at 2°C maintained the viability of the acorns better than unsoaked acorns stored in the same way. In this, his findings agree with those of Suszka and could also be attributed to the enhanced levels of carbon dioxide in the containers.

REFERENCES and FURTHER READING

BONNER, F. T. (1978). Handling and storage of hardwood seeds. *Proceedings of the Second Symposium on Southeastern Hardwoods*, 145–152. USDA Forest Service Atlanta, Georgia, USA.

BONNET-MASIMBERT, M. and MULLER, C. (1975). La conservation des faines est possible. *Revue Forestière Française* **XXVII**, 129–138.

CHIN, H. F. and ROBERTS, E. H. (1980). *Recalcitrant crop seeds*. Tropical Press DN BHD, Kuala Lumpur.

GOSLING, P.G. (1988). The effect of drying *Quercus robur* acorns to different moisture contents, followed by storage, either with or without imbibition. *Forestry* **62**, 41-50.

ROBERTS, E. H. (1972). *Viability of seeds*. Chapman and Hall, London.

ROBERTS, E. H. (1973). Predicting the storage life of seeds. *Seed Science and Technology* **1**, 499–514.

SUSZKA, B. and TYLKOWSKI, T. (1990). Storage of acorns of the English oak (*Quercus robur* L.) over 1–5 winters. *Arb. kornickie* **25**, 199–229.

ZASADA, J. C. and DENSMORE, R. (1980). Alaskan willow and balsam poplar seed viability after 3 years storage. *Tree Planters Notes* (Spring 1980), 9–10.

Appendix 10.1

Storage characteristics of forest tree seeds

Genera (and species)	Moisture content	Temperature	Maximum reported period of storage with little or no loss of germination quality
Conifers			
Abies	8%	1 to 3.5°C	5 years
Araucaria (small-seeded Australasian species)	dry	−15°C	8 years
Cedrus	<10%	−1 to 4°C	3 years
Chamaecyparis	8%	0°C	7 years
Cryptomeria	4-8%	5°C	2 years
Cupressus	dry	1-5°C	20 years
Juniperus	dry	ambient	21 years
Larix	6-8%	1-3°C	25 years
Libocedrus	dry	5°C	2 years
Metasequoia	dry	3°C	>2 years
Picea	4-8%	1-4°C	17 years
Pinus	5-10%	1-5°C	30 years
Pseudotsuga	6-9%	−18°C	20 years
Sciadopitys	<10%	1-5°C	2 years
Sequoia	dry	−18°C	7 years
Sequoiadendron	6-10%	5°C	14 years
Taxodium	dry	5°C	>1 year

Genera (and species)	Moisture content	Temperature	Maximum reported period of storage with little or no loss of germination quality
Taxus	dry	1-3.5°C	6 years
Thuja	5-7%	1-3.5°C	>5 years
Tsuga	8%	−18°C	>5 years
Broadleaves			
Acacia	dry	ambient	17 years
Acer (except A. saccharinum A. pseudoplatanus)	17%	−10°C	4 years
Alnus	airdry	1-3.5°C	10 years
Betula	airdry	4.5°C	12 years
Carpinus	10%	3.5°C	12 years
Fagus	8%	−3°C	5 years
Fraxinus	7-10%	5°C	7 years
Juglans	airdry	2°C	4-5 years
Liquidambar	10%	2-4°	4 years
Liriodendron	10%	2-4°	several
Malus	<11%	2-10°C	>2 years
Platanus	10-15%	−4-4°C	>1 year
Populus	6%	−15°C	6½ years

Genera showing orthodox seed storage characteristics

Genera showing orthodox seed storage characteristics

Genera showing orthodox seed storage characteristics			
Genera (and species)	Moisture content	Temperature	Maximum reported period of storage with little or no loss of germination quality
Prunus	11%	1°C	4½ years
Pyrus	10%	0-5°C	2-3 years
Robinia	airdry	0-4°C	10 years
Salix	6%	<0°C	3 years
Sorbus	6-8%	1-4°C	8 years
Tilia	10-12%	5°C	2-3 years
Ulmus	3-4%	−4°C	15 years

*Dry = airdry in uncontrolled humidity – exact moisture content not calculated.

Genera showing recalcitrant seed storage characteristics.

Acer pseudoplatanus and *A.saccharinum*

Aesculus

Araucaria (large-seeded species: *A.araucana, A.angustifolia, A.hunsteinii* and *A.bidwillii*)

Castanea

Quercus

Chapter 11
Seed Testing

*by **A. G. Gordon***

Introduction

The physical quality of seeds of only 13 species of commercial forest trees is covered by legislation. The species covered are those included in the Forest Reproductive Material Regulations 1977, full details of which have already been given in Chapter 5. All other regulations covering the physical quality of seeds of commercial tree species previously included in the Plant Varieties and Seeds Act 1964, have, since 1973, ceased to have effect. Of the species and hybrid covered by this Bulletin, *Abies grandis, A. procera, Pinus contorta, Larix × eurolepis, Tsuga heterophylla* and *Betula pendula* are not therefore covered by any seed regulations. For these species, normal consumer legislation is all that protects the purchaser.

The Forest Reproductive Material Regulations 1977

Seed of any of the 13 EEC species marketed in Great Britain must be covered by a test certificate issued in respect of it by the Official Seed Testing Station, Forestry Authority Research Station, Alice Holt Lodge, Farnham, Surrey, or in the case of seed imported from a member State, Northern Ireland or a third country, by an officially recognised station of the exporting country. These tests must be made within the seed testing year which begins on 1st July and ends on 30th June in the next calendar year, except that seed marketed during July or August of any seed testing year may be covered by certificates issued in the preceding seed testing year.

Seed marketed under the Forest Reproductive Material Regulations 1977 must be accompanied by a supplier's certificate which must include among others things the following seed testing particulars:

1. number of Test Certificate (if any);
2. the description, 'EEC standard', *or*, a statement that sub-standard seed has been authorised for marketing;
3. percentage of purity;
4. percentage of germination;
5. number per kilogram of live seeds;
6. number per kilogram of seeds capable of germinating;
7. weight of 1000 pure seed in grams;
8. year in which the seed shall have ripened;
9. if the seed has been kept in cold storage.

The EEC standard referred to, concerns the percentage by weight of seed of other forest species present as an impurity in the seed supplied. For conifer species the limit is normally 0.5 per cent and for *Quercus* and *Fagus* 0.1 per cent (see Table 11.1). This characteristic and items 3, 4, 5, 6 and 7 are included in the test certificate issued by the Official Testing Station.

Although other forest tree species are not covered by seed regulations, anyone selling seed is strongly advised to arrange for the physical quality of the seeds to be tested before offering them for sale. This can be done at the Official Testing Station or at private laboratories. For full details of other types of certificates and tests available please consult the companion volume, Forestry Commission Bulletin 59 *Seed manual for ornamental trees and shrubs.*

EEC standard. However, for *Larix* and *Quercus* spp. the limit for seeds of other species of these genera is 1 per cent. This is the only physical standard set in the Regulations on seed testing. Of course it does not apply to non-EEC species.

1000 pure seed weight (seed size)

From the pure seed separated in the purity test eight sets of 100 seeds are counted out at random. These are weighed and provided the sets show sufficiently little variability, they are converted to the 1000 pure seed weight. This is a statistic which takes account of variations in seed size due to origin, weather, etc., and which is used for calculating the number of pure germinable or viable seeds per kilogram. This in turn is used for calculation of sowing densities by nurserymen.

Germination and viability

The tests that are made to assess the ability of a seed lot actually to produce plants are directly related to the level of dormancy in the seed of the species in question. If seeds are deeply dormant (see Chapter 12) requiring very lengthy periods of pretreatment, e.g. *Fagus sylvatica,* then a germination test is impractical. Instead a viability test is carried out. If seeds show no dormancy or possess a shallow dormancy which can be overcome in a short time (no more than 4 weeks), e.g. most common conifers, alders and birches then a germination test is normally given.

It is important for the practitioner to appreciate the difference between viability and germination. *Viability* is the possession in a seed of those processes essential for the seed to germinate. *Germination* is the successful implementation of those processes, leading to the production of a seedling capable of establishment in the nursery. If the conditions given to overcome dormancy (and the conditions during the germination test) are perfect then the germination percentage will match exactly the viability percentage. If on the other hand the pretreatment conditions have been less than perfect, then there will be a discrepancy between the two. In practice it is

seldom possible to achieve perfection in treatment conditions, and the weaker seeds (those possessing less nutrients, less maturity, more processing damage, etc.) will degenerate. The worse the conditions the greater the degeneration and the larger the discrepancy between the original viability figure and the germination achieved. For deeply dormant seed lots it is therefore advisable, if a really accurate sowing density is desired, to test the germination percentage at the end of the pretreatment period. Also, the response of the seed lot to the pretreatment can be monitored during the pretreatment period so that the best moment to end the pretreatment can be identified.

The prescriptions for assessing the germination or viability of tree seeds in the International Rules for seed testing have been based on the above principles. Where a germination test can be carried out in a suitably short time (less than 8 weeks) it has been preferred to the viability test. Viability tests can be and often are given to non-dormant or shallowly dormant seed lots. The most commonly used viability test (by Tetrazolium) is a subjective test, depending much on the experience of the analyst for its accuracy. In theory there should be almost total agreement between results from a viability test of a non-dormant seed lot and from the germination test. In practice this is by no means always the case. Tetrazolium tests are greatly favoured in Germany, the country where the test was developed, and test certificates are frequently issued by that country's official stations based on it, whereas most other countries would use germination tests. The reason given for this practice is that the seed merchant can get his results quicker. As the Tetrazolium results almost always exceed the germination results, the quality may be over-estimated. When interpreting certificates supplied with different German seed lots it is therefore important to appreciate the significance of the method by which the results quoted have been obtained.

Of the species covered by the Forest Reproductive Material Regulations the germination tests of *Quercus* spp. can be completed without difficulty in 4 weeks and a Tetrazolium test is therefore not prescribed by ISTA. However, from experience, some official stations do use the latter test method. Acorns contain chemicals which inhibit the correct reaction with Tetrazolium, and Tetrazolium tests tend therefore to under-estimate the actual germination percentage. For acorns, a reasonably accurate estimate of the viability can be made by simple cutting test at the time of preparation for the germination test. In contrast, results of Tetrazolium tests of *Fagus sylvatica* seeds tend to exceed the actual germination achieved, even after very careful seed treatment.

The prescribed ISTA germination test for *Fagus* seeds lasts up to 26 weeks, which is totally impractical, but experience has shown that a correct interpretation of the Tetrazolium test is only consistently made by experienced seed analysts and on newly collected and vigorous seed lots. As soon as the seed is dried, processed or stored, difficulties in interpretation begin to manifest themselves.

Germination tests

The principle behind the germination prescriptions in the International Rules is the provision of standard conditions so that tests of seed quality made by two different people (for example the seed seller and seed purchaser) can be justifiably compared. Temperature and light conditions are relatively easy to standardise, but moisture availability is not. This is achieved for the great majority of tree seeds by using an inert filter paper medium. The properties of the medium are carefully defined so that it can be copied exactly in different countries. This medium is used for all conifer species covered in this Bulletin plus alders and birches. For the larger *Quercus* seed, the medium is moist sand. Although the size and texture of this is also defined, it is less easy to obtain identical samples in different countries and results can vary somewhat. A filter paper medium is also prescribed for the 26-week long test of *Fagus sylvatica* as it is easier to control the moisture availability using this medium at low temperatures than with sand.

For the majority of tree seeds the prescribed germination temperature and light regimes are 16 hours incubation in the dark at 20°C with 8 hours incubation at 30° in the light. For a large minority, again tending to be the larger seeds incubated in sand, the germination temperature is lower, usually a constant 20°C (see Appendix 6b in FC Bulletin 59). Experience has shown that within quite wide limits these temperatures are not very critical but that sometimes a particular seed lot of a given species does need a lower temperature to achieve maximum germination.

The normal procedure when carrying out a germination test is to select at random four replicates of 100 seeds from the pure seed fraction of seed obtained during the purity test (or if 1000 pure seed weight test has been performed from the eight replicates prepared for that test). These are set up separately on four areas of the germination medium. The size of the seed determines the size of the filter paper or amount of sand that has to be used and may necessitate rearranging the number tested to eight replicates of 50 seeds. Each seed should be separated from each other by at least the width of the seed, to reduce the chance of cross contamination by microflora.

Before the seeds of dormant species are set out, they require some form of pretreatment. Sometimes this is very lengthy, e.g. *Fagus sylvatica*. (It has been pointed out already that the length of the treatment renders the germination test impractical for issuing official certificates on deeply dormant seed lots, although the International Rules do carry such a prescription.) More often the treatment lasts for a shorter period. The normal treatment of most conifers is a prechilling, which involves setting up the seeds in the moisture condition in which they are to germinate, but at a temperature of 3–5°C for 3 or 4 weeks. The large-seeded species, *Fagus* and *Quercus*, must also be prepared by soaking overnight and either cutting a portion of the seed away or peeling off the pericarp.

After the replicates have been treated and

set out, they must be placed like seeds requiring no treatment in some sort of apparatus that controls the temperature, supply of water and light. There are many ways of achieving this, from walk-in room incubators to special cabinet incubators and, the most commonly used in Europe, Jacobson apparatus (an open tank with the seed suspended above the water surface on glass or metal plates). With incubators the moisture may be controlled by maintaining the whole atmosphere at saturation point, by providing moisture to each individual replicate by means of a wick from a reservoir, or by periodically adding more water to the moist sand. With the Jacobson apparatus (also called the Copenhagen tank) the water, which itself controls the temperature, is supplied direct from the tank by means of filter paper or cotton wicks.

For seeds of all conifers, *Betula pendula* and *Quercus* spp., the incubation period is from 21–28 days. After 7 days, seedlings which have developed well enough to convince the analyst that they would have grown into healthy plants in the field are identified, removed and counted. This is repeated weekly until the end of the test. These so-called normal seedlings are differentiated from abnormal seedlings (those deemed to be unable to develop into healthy plants in the field) which are also counted. To aid analysts in the interpretation both normal and abnormal seedlings are defined by species or genus in the International Rules. Further amplification is given in the *Handbook for seedling evaluation* (Bekendam and Grob, 1979). A handbook dealing exclusively with tree species has been published by ISTA (Gordon, Gosling and Wang, 1991).

When the period of incubation is complete, the ungerminated seeds should be cut open to see if any are still fresh and healthy looking. These are the seeds which are still alive, but which have not reacted to the treatment conditions by germinating. They are still in a dormant condition and are called fresh ungerminated seeds. They can be distinguished with more certainty by using the Tetrazolium test (see below). Other ungerminated seeds will be dead and will have decayed to varying degrees during the course of the test. Others will be empty or nearly empty. The percentage of these is a useful characteristic for a seed merchant to know, as it gives him some indication of how well the seed has been processed. This characteristic may be reported upon request on ISTA certificates. The empty seeds counted at the end of a germination test will include a proportion which have become so because of the decomposition of their contents during the course of the test. A more accurate assessment of truly empty seed can be made therefore on a separate sample before the start of the test. Another alternative method, recently included in International Rules, is to assess the proportion of empty seed non-destructively by X-raying the replicates to be germinated before the start of the germination test.

The numbers of normal and abnormal seedlings and dead, empty and fresh seeds are added up for each replicate and should equal one hundred. If the variability between the number of normal germinants for the four replicates is not too great (special tables are available in the International Rules to test this) the average of the four results is taken as an expression of the quality of the seed lot. If the variability is excessive it indicates it has been caused by something other than the differences that could be expected by random sampling. For official or international tests which exceed the tolerance limit, it is necessary to repeat the whole test procedure. The average germination percentages of the species covered in this Bulletin recorded over the last 30 years at Alice Holt Lodge are listed along with seed weight data in Table 11.4.

Viability tests

There are two main types of viability test, the staining tests and the excised embryo test. In the former the viability of the seeds is assessed by their ability to change the colour of liquid (e.g. Tetrazolium or indigo carmine). The assessment of staining tests is difficult to carry out and for this reason the excised embryo test which is a form of growth test is sometimes

Table 11.4 Seed quality of commercial forest species based on 30 years of seed test results at Alice Holt Lodge

Species and common name	Average number of pure seeds per kg	Average laboratory germination percentage or viability (v) %	Average number of germinable seeds per kg (a)
Conifers			
Abies grandis (Grand fir)	45,000	40	20,000
Abies procera (Noble fir)	40,000	35	15,000
Chamaecyparis lawsoniana (b) (Lawson cypress)	460,000	50	230,000
Larix decidua (European larch)	170,000	40	70,000
Larix leptolepis (Japanese larch)	250,000	40	100,000
Larix × *eurolepis* (Hybrid larch)	210,000	30	60,000
Picea abies (Norway spruce)	145,000	80	110,000
Picea sitchensis (Sitka spruce)			
Home collected	400,000	90	360,000
Imported	400,000	60	240,000
Pinus contorta (Lodgepole pine)	300,000	90	270,000
Pinus nigra var. *maritima* (Corsican pine)	70,000	80	55,000
Pinus sylvestris (c) (Scots pine)	165,000	85	140,000
Pseudotsuga menziesii (Douglas fir)	88,000	80	70,000
Thuja plicata (b) (Western red cedar)	850,000	60	500,000
Tsuga heterophylla (b) (Western hemlock)	650,000	65	375,000
Broadleaves			
Betula pendula (Silver birch)	1,900,000	25	150,000
Fagus sylvatica (Beech)	4,600	70(v)	3,000(v)
Quercus petraea (Sessile oak)	315	57	180
Quercus robur (Pedunculate oak)	270	72	195
Quercus rubra (Red oak)	280	71	200

Notes: a. The figures included in this column are the number of germinable seeds that can be expected in a kilogram of each species. For further explanation of this concept see Table 23b in *Nursery practice* FC Bulletin 43 (1972), or its equivalent later revision.

b. These species were included in Table 23b of *Nursery practice* and have been included here for the sake of completeness.

c. Although not conclusive a tendency has been noted for seed of *Pinus sylvestris* from clonal seed orchards to be larger in size and hence the number of germinable seeds per kilogram to be lower than seed produced from general collections in forest situations.

preferred. This involves the incubation of embryos, carefully removed from their seed coverings. Both the Tetrazolium test and the excised embryo test are described in detail in the International Seed Testing Rules. The only species covered in this manual for which a viability test is officially prescribed is *Fagus sylvatica*.

When official tests are performed, sampling should be carried out in exactly the same way as for germination tests. However, preparing and assessing four replicates of 100 seeds for either of these viability tests is extremely time-consuming and laborious, and for unofficial tests the number of seeds treated is often reduced to four replicates of 50. The same tolerance limits are used for the full Tetrazolium and excised embryo tests as for germination tests.

Tetrazolium tests
The principle on which the Tetrazolium test is based is that active dehydrogenase enzymes in the seed react with the colourless solution of 2,3,5–triphenyltetrazolium bromide or chloride to turn it bright red. The distribution of the insoluble red pigment in the important structures of the embryo and ensdosperm,

gives an indication if the seed is viable. Detailed instructions are given in the International Rules as to how the 1 per cent buffered tetrazolium solution and the seeds should be prepared and how the embryos should be evaluated. The cost of the salt at 1991 prices is £12 per 10 g.

The aim of the seed preparation is to expose the embryo and endosperm to the Tetrazolium during staining without damaging them in critical places. The exact method of preparation varies from species to species, but for *Fagus sylvatica* it involves soaking the seeds in water before removing the pericarp and seed coat. The seeds are then placed in the Tetrazolium solution for 24 or more hours at a temperature of 30°C in complete darkness (the solution changes colour in light). At the end of this period the solution is decanted, and after the seeds are washed in water each one is assessed as viable or non-viable. During the assessment the embryos should be kept moist. The acceptable staining pattern for each species is given in the Rules, but in essence, the vital areas where the cell divisions take place (the root tip and the plumule) and where the cotyledons are attached to the main axis should be fully stained. For seeds with endosperm, this tissue must also be well enough stained to provide nutrient for satisfactory seedling growth. In most species this means full staining.

In order to report the result of a Tetrazolium test on an official ISTA certificate the ISTA prescription must be followed exactly. However, for advisory purposes, alterations to the dissection technique, changes in Tetrazolium concentration, incubation temperatures and duration can make the test simpler and quicker without affecting the results too markedly. A variation which reduces the preparation time significantly is that used widely in America. Seeds are sliced longitudinally through the embryo (and endosperm) and placed in solution of lower concentration 0.1 to 0.5 per cent for shorter periods. By examining the staining pattern under a binocular microscope an estimate of the viability can be made which has been shown to agree closely with the classical method. When used in this way Tetrazolium testing can be a very useful and versatile tool of the greatest possible help in monitoring the quality of seed during processing. However, being a very subjective test, it requires considerable experience to ensure meaningful evaluation. This can only be gained by practice. Unfortunately the 1976 and previous ISTA prescriptions have not given uniform results between laboratories. They have been completely rewritten and published in the new version of the International Rules (ISTA, 1985). A complete description of the technique including much additional practical detail can be found in the *Tetrazolium handbook* (Moore, 1985).

The Tetrazolium test gives an estimate of the viability of the seed at the time of the test. For species with little or no dormancy this will agree quite closely with the actual germination test results. However, for very dormant species like *Fagus sylvatica* the relationship may be less close. This is because the number that germinate will be determined by the pretreatment and germination conditions to which the seed is subjected. If these are perfect then most if not all the seeds assessed as viable will actually germinate. If the conditions are unfavourable then weaker seeds, e.g. those with damaged parts, will not germinate. For this deeply dormant species it is necessary to classify only those seeds stained perfectly if a meaningful estimate of the future germination quality is to be made. For further discussion on this subject refer to the 'Germination and viability' section on pp.108–109.

Other staining tests

Many other chemicals possess the ability to react to dehydrogenase enzyme systems. A large number have been tried in the past. The only other compound still used widely is indigo carmine. Test certificates issued in eastern European countries may well report this test. It differs fundamentally from Tetrazolium in that live tissues remain unstained, while dead tissues turn blue. In this test only the embryos are examined. Correlations have been obtained with germination that are comparable with those of Tetrazolium tests. The chemical is much cheaper, more uniform between manufacturers products and less toxic to operators than Tetrazolium, and with experience can give very satisfactory results (Rostovtsev and Lyubich, 1978).

Excised embryo test

The principle on which the excised embryo test is based is that viable embryos dissected from seeds will remain healthy and often begin to grow during a period of incubation, whereas dead, dying and even some weak embryos will decay during the same incubation period.

Detailed suggestions for preparation of the seeds are given in Appendix C of the International Rules. They were introduced in 1976 on a provisional (non-prescription) basis. They vary from species to species and differ in some important respects from the seed preparation prescriptions for the Tetrazolium test of the same species. Most include a period of soaking often following a treatment to allow the water to penetrate. All methods are designed to expose the embryo with as little damage as possible.

Although the excised embryo test is not prescribed officially for any commercial tree species, it can be used, as is the Tetrazolium test, instead of the germination test for testing the viability of almost any species. On occasions laboratories may report viability using this method.

Accurate calculation of seed quality

From the three tests described above the following characteristics are obtained:

the percentage purity=P
the weight of 1000 pure seeds=W in grams
the germination or viability percentage=G (or V).

The number of germinable or viable seeds in a kilogram of the bulk is calculated thus:

$$\frac{1000 \times 1000}{W} \times G \text{ (or V)} \times P$$

example

for *Pinus sylvestris* Purity   =99.8%
                        1000 psw   =7.699g
                        Germination =91%

$$\frac{1000 \times 1000}{7.699} \times \frac{91}{100} \times \frac{99.8}{100} = 117,961$$

This is normally expressed to the nearest thousand, i.e. 118,000 germinable seeds per kilogram.

Other tests

Moisture content

The moisture content of tree seeds is often a good indicator of its germination quality. This is because of the correlation that exists between the measured moisture content and the ability of seeds to store safely.

The normal method of determining the moisture content of tree seeds is by an oven method. Seeds are dried for 16–48 hours in an oven set at 105°C. The loss in weight, which is interpreted as loss of moisture, is expressed as a percentage of the fresh seed weight at the start. For certain conifer species with high resin and fat levels in their seeds heating will drive them off. This will give figures in excess of their true moisture content values. The only way to obtain the true moisture content of such seed is to use very time-consuming toluene distillation methods or the more recently developed Karl Fischer reagent test. In practice in Europe these are very rarely used. Agricultural seed moisture meters using electrical conductivity have been calibrated for forest seeds and have been found to give reasonably accurate results, good enough for practical decisions on storage moisture contents.

Seed health

Many countries outside Europe require documentary evidence that any tree seed imported does not suffer from harmful organisms, either fungi or insects. This evidence is supplied by means of a plant health (phytosanitary) certificate. To obtain a plant health certificate in Britain, the seed lot must be sampled by the local plant health and seed inspector (PHSI) of the Ministry of Agriculture, Fisheries and Food. The sample taken is forwarded to the MAFF Plant Health Laboratory at Harpenden, from where it may be sent to Alice Holt Lodge for detailed examination by the Principal Pathologist.

The seed (both dry and during a germination test) is examined for the presence of fungi known to be pathogenic to germination and growth of the seedlings. It is also examined for the presence of insects. The best method of detecting insect contamination is by X-ray. Failing this the seeds can be cut open and physically examined. If the sample is found to be free of pathogenic fungi and insects, a recommendation for certification is forwarded to the Certifying Office in London who issue the necessary certificate. Due to the many stages involved, at least 3 weeks should be allowed between sampling and receipt of the certificate. Anyone wishing to export seeds to a country outside the EEC should consult the local PHSI to check the specific regulations of the importing country. Many countries require the seeds to be fumigated or given a special treatment as part of the plant health procedure.

In practice very few seed-borne fungal pathogens of forest tree seeds have been identified and Britain already has endemic populations of the normal seed infecting insects. With few exceptions (*Quercus, Castanea* and *Prunus* spp.) Britain has no phytosanitary restrictions on tree seeds. The national regulations on seed health are included in the Plant Health Regulations which have been introduced in line with the Plant Health Directive of the EEC (see Statutory Instruments Nos 420, 449, 450 and 499, 1980; 807, 1983; 1892, 1984; 196, 1986; 1758, 1987 and 1951, 1989). See Phytosanitary requirements p.53 in Chapter 5.

Heterogeneity tests

When conifer seed is stored in bulk the empty seeds tend to settle out on the surface. Unless great care is taken a quantity of seed withdrawn from the bulk may not be representative of the whole. For this reason seed is normally stored in relatively small lots at Alice Holt Lodge and in order to be certain

that the component parts of a seed lot are evenly distributed, a heterogeneity test is performed. Because of the legal implications, persons selling large lots of seeds with high empty seed contents in large numbers of different sub-lots are strongly advised to have the component parts in which the seeds are stored tested for homogeneity. A good indication of the degree of homogeneity can be obtained locally by randomly sampling each component of the bulk, cutting at least 100 seeds with a sharp blade and assessing the number of empty seeds. A simple calculation, details of which are contained in Chapter 2A of ISTA Rules will indicate if the empty seeds are randomly distributed.

Quick information test

For species not covered by the FRM Regulations, and for seed collected by the nurseryman himself, a quick information test instead of a full statutory test may be all that is necessary. For as little as £17.50 (1991 prices) an estimate of quality per unit weight is given by the Official Testing Station (usually within 10 days) based on reduced test samples for broadleaved species. Samples submitted for this test should also be sent in thick paper envelopes to the Officer-in-Charge clearly marked Quick Information Tests.

Quality and sowing density

The practical use to which measurement of seed quality is put is the calculation of a suitable sowing density for individual seed lots. This is essential if valuable seed or valuable nursery space is not to be wasted. At the end of germination or viability tests a figure is available which is the estimate of the number of germinable or viable seeds in a kilogram of the product sold. This is to be found on the Seed Suppliers Certificate (see Chapter 5) or on the Seed Test Certificate if it has been requested. The estimate is based upon the number of seeds that germinated in ideal conditions and makes no attempt to estimate the number that will germinate in the nursery. To a limited extent this will be determined by the vigour of the seed (which is not measured or reported in official tests – new season's seed tend on the whole to be more vigorous than stored seeds) but to a much larger extent the number of seedlings produced will be governed by dormancy and the environmental factors into which the seed is sown. Full details of the calculation of sowing density and the way this is affected by seed quality and the expected yields are to be found in Forestry Commission Bulletin 43 *Nursery practice* (1972); this publication is out-of-print, but a new edition is in preparation.

The average quality of seed as measured by the tests described above is given in Table 11.3. The figures differ in certain minor respects from those given in Table 23b of Bulletin 43 (1972). This is because they have been updated to include more recent data from the tests carried out by the Official Seed Testing Station at Alice Holt Lodge, Farnham, Surrey, since 1972. It is interesting to note a slight increase in some germination percentages and the very significant difference in the germination quality of imported as opposed to home produced *Picea sitchensis*.

REFERENCES

BEKENDAM, J. and GROB, R. (1979). *Handbook for seedling evaluation.* International Seed Testing Association, Zurich, Switzerland.

GORDON, A.G., GOSLING, P. and WANG, B.S.P. (1991). *Tree and shrub seed handbook.* International Seed Testing Association, Zurich, Switzerland.

ISTA (1985). International rules for seed testing 1985. *Seed Science and Technology* **13** (2), 299–355.

MOORE, R.P. (ed.) (1985). *Handbook on tetrazolium testing.* International Seed Testing Association, Zurich, Switzerland.

ROSTOVTSEV, S. A. and LYUBICH, E.S. (1978). Determination of the viability of tree and shrub seeds by staining with indigo carmine in the USSR. *Seed Science and Technology* **6**(3), 869–875.

Chapter 12

Seed Dormancy, Seed Treatment and Seed Sowing

by **A. G. Gordon**

Introduction

Mature seeds of most woody plant species from the temperate zones will not germinate promptly when placed under conditions which are normally regarded as suitable for germination. Such seeds are said to be dormant.

Seed dormancy aids the survival of wild plant populations by preventing out-of-season emergence, and by spreading the germination over many weeks or even years, thereby increasing the opportunities for successful seedling establishment. In contrast, the same phenomenon presents a problem to the nursery manager who is attempting to produce large, uniform crops of seedlings as rapidly as possible. However, the problem is not insurmountable and dormant seeds can be induced to germinate within a reasonable time if certain predetermined conditions are satisfied.

The mechanisms which restrict germination vary widely, so dormancy is a rather vague and relative term. It describes a condition which results both from inherited properties and from conditions imposed by the environment. It may vary in degree within a species, depending on differences between individuals (as Gordon, 1978, showed for *Picea sitchensis* and *Pseudotsuga menziesii*), between locations (ISTA, 1976, for *Pinus sylvestris*), climatic conditions or times of collection and it is certainly modified by different methods of processing and by nature and duration of storage.

Attempts have been made to put the subject into some kind of order but there is still no generally accepted classification. A thorough description of various schemes is included in the companion volume, Forestry Commission Bulletin 59 *Seed manual for ornamental trees and shrubs*. A system that was designed specifically with trees in mind was that of Nikolaeva (1967; 1977). This was described in some detail in the above Bulletin.

Seed dormancy in conifers

Although some conifers exhibit a deep dormancy, caused mainly by a hard seedcoat (mechanical dormancy), and *Pinus peuce* exhibts embryo dormancy, the great majority of temperate conifers exhibit a weak physiological dormancy. This normally results in a proportion of the seeds in a lot failing to germinate in the normal period for germination, and therefore in an uneven seedbed at the end of the growing season. Keeping seeds moist at a low temperature for several weeks has been found to increase the uniformity of germination of coniferous seed which shows this weak physiological dormancy. This moist prechilling treatment in fact improves both the rate of germination and the total seedling yield of all normal conifers even when they have not been dormant (Buszewicz and Gordon, 1974; 1975; Gordon, 1976). This has been found to be particularly true for *Thuja plicata* and *Chamaecyparis lawsoniana*. When tested in the laboratory these species often show no reaction in their germination percentage to prechilling and yet the yield of seedlings in the nursery may be dramatically improved by a cold moist treatment (Gordon and Wakeman, 1979). Most conifer species mentioned in this Manual show some physiological dormancy from time to time but *Picea sitchensis*, *Larix leptolepis*, *Abies grandis*, *Tsuga heterophylla* and *Pinus contorta* (in descending order of incidence) show it more frequently than other species.

Conifer seed treatment

It has been found that moist prechilling treatments of conifer seed are more efficiently carried out in transplant cold stores or even domestic refrigerators than in stratification pits in the nursery, where the temperature cannot be accurately controlled (Gordon, 1976). It has also been shown quite clearly that for all nine coniferous species studied so far, 'naked' pretreatment in the absence of sand is almost always better and never worse than pretreatment in sand.

The following procedure should be followed for all coniferous seed unless evidence is obtained from the seed test that the seed lot in question suffers damage from cold moist conditions (see below).

The bulk of the seed lot should be divided up into quantities corresponding to particular seedbed areas in order that the seed can be later sown at the correct sowing density. These seed lots should then be placed into suitable containers (thick polyethylene bags are ideal) and several times the volume of cold water added. The soaking seed should then be placed in refrigeration at approximately 3–5°C for 48 hours making sure that the seeds do not float and remain dry. The seeds are then drained of water until only the very bottom seeds are standing in water. This can easily be done by puncturing the bottom corners of the bags with a knife and hanging them up so that the water drains through the holes. The perforated bag and seeds should then be placed inside another bag and the whole placed in refrigeration at 1 to 5°C. Once a week the bags should be opened, the seeds mixed and remoistened with a water spray if the surfaces show signs of drying out. In order to give flexibility for sowing plans it is best to aim to leave the seeds in the refrigeration for 4 weeks before sowing, but

with a flexibility of −1 or +2 weeks so that seeds can be sown any time over a period of 3 weeks after the basic initial 3 weeks treatment whenever seedbed conditions are good enough. Seed at this point will have a moisture content of about 30 per cent. Much longer periods of chilling for up to 12 weeks at 0.5°C have recently been found to be particularly beneficial in obtaining uniform and very complete germination. Prechilling at this temperature does not produce the incipient germination after 9 weeks found when temperatures of 3 to 5°C have been used.

At the end of the pretreatment period the seed, still in lots for sowing, should be spread thinly in trays and allowed to surface dry without artificial heat in a cool, well ventilated area. It has been shown that the benefit of pretreatment is not lost when seed is dried down to 25 per cent moisture content and kept there for several days. At this moisture content the seed just flows freely and can be sown in a mechanical seed sower. When it has reached this condition, before final sowing and at any time during the final seed preparation, if seedbed conditions deteriorate the pretreated seed should be replaced in refrigeration in this surface-dried state in lightly-tied polythene bags (Buszewicz and Gordon, 1974).

Root emergence will not take place provided these recommended pretreatment periods are not exceeded and the temperature does not rise above 6°C. If rootlets are observed on many seeds the seed lot should be sown as quickly as possible.

Priming of seeds with polyethylene glycol (PEG) is a technique that has been tried for many horticultural species with considerable success in order to produce a more instantaneous and synchronous germination. Trials with the main conifer species used in Britain failed to produce a stimulus that could not also be produced as well and much more easily by the standard prechilling regimes described above.

One technique has been developed for conifers in North America which seems to go rather beyond the normal effect of chilling. It is known as seed conditioning and was developed first with *Abies* spp. (Edwards, 1981). Seed is moist chilled in a way similar to that described above, for a period of approximately 6 weeks. It is then partly dried back from about 30–35 per cent to around 20 per cent and kept for several months at 1°C. At the end of this period more seed is found to have germinated than had previously been found to have been viable. One explanation of the phenomenon is that the long period at low temperatures (and high oxygen pressures) has led to repair of membranes in cells of old seeds that had previously degenerated beyond the point of exhibiting viable symptoms. This technique offers interesting possibilities for improving the vigour of seed lots that had begun to deteriorate through age.

Research experience has shown that early sowing of seeds as a substitute for seed pretreatment cannot be guaranteed to supply the pretreatment required. At times and in certain locations it can adequately substitute but it should not be risked if refrigeration is readily available. Research has also shown that some seed lots of *Picea sitchensis* imported from Canada carry a small proportion of seeds infected by a fungus *Geniculodendron pyriforme* which can spread when the seeds encounter cold moist conditions. These can be found in early sown beds and in prechilling treatments. Contamination as low as one seed in 1000 can result in reductions in germination of 30 per cent over seed sown without chilling treatment (Gordon, Salt and Brown, 1976). The problem is now recognised and steps can be taken to remove the contamination. No British produced seed has ever been found to be contaminated.

Occasionally a seed lot of other species is found to react to the standard chilling treatment by a reduction in germination. This seldom reaches a high level and is not associated with *G. pyriforme*. If the reduction in absolute germination on such lots is excessively high, it is best not to prechill the seeds but where the reduction is less than 5 per cent the increase in rate of germination from chilling more than compensates for the reduction in absolute numbers as expressed by the number

of usable seedlings at the end of the season.

Seed dormancy in broadleaves

There are many different forms of dormancy exhibited by broadleaved species but of the commercially important ones, only *Fagus sylvatica* poses any real germination problems. *Betula pendula* exhibits weak physiological dormancy and reacts positively to moist chilling treatments as described above for conifers. Of the three *Quercus* spp. only *Q. rubra* exhibits deep radicle dormancy, in common with other species in the red oak group. When sown in the autumn the dormancy is readily overcome by the cold winter temperatures. If stored for spring sowing, the dormancy disappears naturally due to the cooler winter temperatures. Seed of *Quercus* spp. in the white oak group germinate very readily and, in fact, it is often very difficult to prevent acorns from germinating during collection and processing. In this respect *Q. petraea* seems more prone to pre-germination than seeds of *Q. robur*. Provided the warmth and moisture given off in respiration during pre-germination does not lead to 'heating', little adverse effect results from these sprouted acorns provided the roots are not allowed to grow too long. Acorns in this condition are therefore best sown as soon as possible after receipt. Sometimes sprouted acorns will lose their sprouts when moved around in bulk and this may even prove advantageous as the resulting seedling fails to develop the normal strong tap root. These two *Quercus* spp. do, however, exhibit an epicotyl dormancy at times. This is a dormancy of the embryonic shoot after full radicle emergence and growth and is only overcome by a period of cold in the winter following radicle emergence.

There is a wide variation in the degree of dormancy within different seed lots of *Fagus sylvatica*. To some extent this is caused by the various methods of preparing and storing the seeds. Thus seed dried to a low moisture content and stored below zero will not lose any dormancy, whereas seeds stored at a high moisture content at just above zero will in fact undergo a kind of moist chilling treatment.

Thus they will not have such a deep dormancy. However, it has also been shown that seed from different localities treated in exactly the same way will show quite wide variations in their dormancy. If germination is to be successful, each individual seed lot must receive special treatment. In continental countries where much *F. sylvatica* seed is sown, special tests have been devised to determine the amount of chilling treatment required.

Unfortunately, in Britain it is often impossible to carry out these tests since it is usually found that the treatment should have begun for the bulk before the results of the sample test have been completed. It is for these reasons that *F. sylvatica* seeds are often sown in the autumn so that dormancy may be overcome naturally as a result of the low temperatures throughout winter. However, such a practice should only be undertaken if it is possible to protect the seedbeds adequately from vermin, birds and spring frost following early germination. This may mean surrounding the whole area with mouse netting and protecting each seedbed with nylon nets. Areas prone to flooding in late winter and spring should be avoided since *F. sylvatica* is inclined to rot under these conditions.

The majority of *F. sylvatica* seed is therefore sown in the spring after seed treatment. But the great difficulty is to know just how long the seed should be treated to ensure maximum germination without running the risk of the germinating seedlings being caught by late spring frosts. If the seed is available early enough it is possible to carry out a pilot test to determine how long the period of treatment should be. *F. sylvatica* is a species that will germinate slowly but quite readily at normal prechilling temperatures. By soaking a sample and storing it at 1–5°C and observing the emergence weekly, the length of the chilling measured for the bulk can be quite accurately determined. Various methods of moistening *Fagus* seeds have been practised over the years, including spraying them and mixing them in layers with moist compost or soil. However, a technique similar to that described for conifer treatment above has given excellent results on both small and large scales. The aim of all techniques is to raise the moisture content to 30–35 per cent and to keep the seeds in this moist condition with adequate aeration at a temperature near 3°C. No matter the treatment given it is advisable to count the number of seeds that germinate each week in order that some idea of the loss of dormancy can be obtained.

On average the length of treatment required is 12 weeks but this can vary from 4 to 16. It is therefore suggested that treatment should begin on the assumption that a 12 week period is required unless a test carried out previously has shown otherwise. The chilling treatment should continue until about 50 per cent of the potential germinable seeds have begun to show signs of radicle emergence. At this time the seed should be carefully sown at the recommended density in properly prepared seedbeds. If, however, this condition is reached at a time when it is inadvisable to sow because of the danger of late spring frosts, the growth of the emerged roots must be brought to a stop. This can be done very successfully by putting the seeds in a temperature below 0°C. It has been found that seeds of this sort can tolerate temperatures down to −6°C and it is a practice commonly employed in continental European nurseries. In these conditions the growth of the radicle will stop almost completely and the seed can be kept there quite safely for several weeks. The shorter the storage period the better, but some batches have survived with little damage for up to 12 weeks. It is advisable before sowing to raise the temperature slowly to avoid the risk of damage to the fragile radicles.

In some seed lots the seeds may begin to germinate over a very wide range of times. This makes it very difficult to determine exactly when the treatment should stop. If the treatment is continued the early germinating seeds will produce excessively long radicles and the subsequent seedlings will fail to survive. Such a seed lot may well be one produced by mixing seed from different sources. However, if faced with the problem it is advisable to attempt to remove the first germinating seeds by flotation and sieving and to return the balance to the cold treatment. In extreme cases separation of the chitted seeds may be necessary on more than one occasion.

Seed sowing

A lot of hard work and expensive seed can be wasted if successful germination does not follow good seed treatment. Most aspects of seedling production are more appropriate to a nursery practice manual (see Forestry Commission Bulletin 43 *Nursery practice* (1972) or later revision), but some of fundamental importance to seed germination are appropriate to a seed manual.

Intimate contact between seed and seedbed

Roughing up the surface of rolled seedbeds prior to sowing, to ensure intimate contact between seed and seedbed, resulted in significant improvements in the rate and percentage germination over seed sown directly on to the flat surface. The importance of sowing fine seeded species on the surface of seedbeds and not under the surface was clearly demonstrated (Buszewicz and Gordon, 1974; Gordon, 1977; Gordon and Wakeman, 1978; 1979; 1980).

Seedbed coverings

Comparisons of different sands and grits showed that in the main small-seeded species germinated better after covering with sand than after covering with grit. In years of drought however grit gave higher germinations, particularly when applied at double the normal rates. More recently it has been found that small variations in the proportions of small sand-sized particles in a grit can produce significant differences in germination percentages of small-seeded species (Gosling, 1983; 1984). Temperatures were recorded under the surface of grits and sands, and these recordings showed how important the colour of seedbed coverings was in keeping down soil temperatures at the level

of the germinating seed. White sand and grit kept seedbed temperatures to a minimum but a buff-coloured builders' sand resulted in temperatures in continuous sunshine in May of 30°C plus, easily capable of sending seeds into thermo-induced dormancy (Gordon and Wakeman, 1979; 1980).

Soil moisture deficits

Examination of meteorological data over a 10-year period for 12 nurseries located throughout Britain showed that in every year, without fail at some period of the germination and growing season, soil moisture deficits in excess of 6 mm of moisture built up and thus put germination and survival at risk. Irrigation experiments produced consistent and highly significant increases in germination of every species tested even in years of average rainfall. Sowing prechilled seeds into a dry seedbed in the drought year of 1976 gave significantly higher germination than sowing untreated seed but application of irrigation raised the germination from prechilled seeds to near normal levels. The absolute requirement of *Betula* seeds for high levels of seedbed moisture was clearly demonstrated (Gordon, 1976; 1977; Gordon and Wakeman, 1978; 1979; 1980).

Seedbed protection

Investigations of the fate of known numbers of seeds sown on to standard nursery seedbeds revealed that a significant number of seeds were lost within days of sowing. The majority were eaten by birds. The worst predation occurred in nurseries on the flight path of migratory birds. Successful protection was afforded by fine plastic netting suspended over wire hoops and tightly anchored to the ground. Removal of such netting had to be delayed until after the seed remnants had fallen from the seedling; birds frequently ate such remnants thereby almost always damaging the growing point beyond recovery. Covering seed with a noxious seed dressing did not reduce the predation significantly when the bird population was very high (Buszewicz and Gordon, 1972; 1973; 1974).

REFERENCES

ALDHOUS, J.R. (1972). *Nursery practice.* Forestry Commission, Bulletin 43. HMSO, London. *(Revision in preparation.)*

BUSZEWICZ, G.M. and GORDON, A.G. (1972). *Report on forest research 1972,* 25–26. HMSO, London.

BUSZEWICZ, G.M. and GORDON, A.G. (1973). *Report on forest research 1973,* 28–30. HMSO, London.

BUSZEWICZ, G.M. and GORDON, A.G. (1974). *Report on forest research 1974, 9–10. HMSO, London*

BUSZEWICZ, G.M. and GORDON, A.G. (1975). *Report on forest research 1975, 7–9.* HMSO, London.

EDWARDS, D. G. W. (1981). Improving seed germination in *Abies. International Plant Propagators' Society, Combined Proceedings* **31,** 69–78.

GORDON, A.G. (1976). *Report on forest research 1976,* 7–9. HMSO, London.

GORDON, A.G. (1977). *Report on forest research 1977,* 7–9. HMSO, London.

GORDON, A.G., SALT, G.A., BROWN, R.M. (1976). Effect of pre-sowing moist chilling treatments on seed bed emergence of Sitka spruce seed infected by *Geniculodendron pyriforme* Salt. *Forestry* **49** (2), 143–151.

GORDON, A.G. and WAKEMAN, D.C (1978). *Report on forest research 1978,* 7–8. HMSO, London.

GORDON, A.G. and WAKEMAN, D.C. (1979). *Report on forest research 1979,* 8–9. HMSO, London.

GORDON, A.G. and WAKEMAN, D.C. (1980). *Report on forest research 1980,* 8–9. HMSO, London.

GOSLING, P.G. (1983). *Report on forest research 1983,* 8–9. HMSO, London.

GOSLING, P.G. (1984). *Report on forest research 1984,* 8–9. HMSO, London.

NIKOLAEVA, M.G. (1967). *Physiology of deep dormancy in seeds.* Israel Program for Scientific Translations, Jerusalem. (1969). 220 pp.

NICOLAEVA, M.G. (1977). Factors controlling the seed dormancy pattern. In, *Physiology and biochemistry of seed dormancy and germination,* ed. A.A. Khan, 51–74. Elsevier, Holland.

Glossary

Achene a small dry indehiscent fruit

Almost Plus seed stand a stand of trees in which between 50 and 75% of the dominant trees are healthy and have straight stems and good well shaped crowns. This term was used prior to the introduction of the Forest Reproductive Material Regulations in 1973.

Angiosperm woody or herbaceous plants: ovules completely enclosed in an ovary which is usually crowned by a style and stigma. Microspores (pollen grains) adhere to stigma; fertilization effected by means of a pollen tube; xylem containing vessels.

Anther the part of the stamen containing pollen grains.

Antheridium structure containing the male sexual cells.

Apices growing points, or zones of cell division, at the tips of stems and roots in vascular plants.

Archegonium the structure in the flower, containing the female sexual cell.

Artificial pollination system where pollen is collected from male flowers artificially and applied to female flowers at the receptive stage sometimes after long-term cold storage.

Axil the junction of leaf and stem at the position nearest the tip of the stem.

Axillary bud a bud in that position.

Basic material (under Forest Reproductive Material Regulations) In relation to forest reproductive material
(a) produced by sexual means, stands of trees and seed orchards;
(b) produced by vegetative means, clones and mixtures of clones.

Berry a fleshy fruit, usually several seeded, without a stoney layer surrounding the seeds.

Bract a modified leaf which extends underneath a scale in female cones.

Bushel old volumetric measure for cones. 1 bushel = 8 gallons = 36.48 litres.

Canopy closure when branches of trees in juvenile stand first overlap and begin to restrict light and vegetation on forest floor.

Caseharden inability of cone scales to reflex, caused by rapid drying of the outer layers while the inner layers remain moist, thereby preventing the further exit of moisture.

Cataphyll early leaf form of a plant, e.g. cotyledon, bud scales.

Catkin see Strobilus.

Certificate of Provenance certificate issued by the authority appointed by National governments, describing the provenance of seed or plants.

Clonal orchard see Seed orchard

Clone a population of genetically identical plants produced vegetatively from one original seedling or stock (and may include parts of plants so produced).

Cone the reproductive structure of conifers (see also Pollen cone, Seed cone).

Cotyledons primary leaves of an embryo or seedling which degenerate soon after the plant produces the first true leaves.

Cultivar plant variety, resulting from breeding, and/or selection and/or cultivation; conventionally denoted by species name, followed by "cv" and cultivar name.

Dehiscent opening when dry to shed seed or spores.

Derogation term used in dealings with the European Economic Community to describe the granting of permission by the European Commission to market goods or perform some action within a member country which is not normally permitted within the rest of the Community.

Dioecious having sexes on different plants.

Diploid an organism or cell having double the basic or haploid number of chromosomes (usually abbreviated as 2N), characteristics of almost all vascular plant cells except gametes.

Dominant (trees) the tallest, largest-crowned trees in closed canopy woodland.

Dormancy a physiological state in which a seed that is capable of germination does not germinate, even in the presence of favourable environmental conditions.

Drupe a more or less fleshy fruit, with usually one (sometimes more) seeds each surrounded by a stoney layer.

EEC Standard seeds of the species covered by the Forest Reproductive Material Regulations which contain less then a prescribed quantity of seed of other forest species or voluntarily, plants which attain certain standards as to height and stem diameter.

Embryo the product of fusion of a male gamete with an ovule during fertilization (2N). In conifers, the embryo is enclosed by storage tissue and the seedcoat, and under favourable conditions grows into a new plant.

Empty seed a seed that does not contain all tissues essential for germination.

Endosperm a commonly used, but incorrect, term applied to the nutrient storage tissue (1N) surrounding the embryo in gymnosperm seeds. This tissue, which is the megametophyte, serves the same function as the endosperm of angiosperm seeds. However, fertilization is not required for this tissue to form in gymnosperm seeds.

Family all plants derived from the same cross between male and female parents.

Female gametophyte see Megagametophyte

Fertilization penetration of a pollen tube into the ovule and union of the male and female nucleii.

Filled seed a seed containing all tissues essential for germination. (Also used in cone evaluation to describe a seed containing storage tissue, but not necessarily an embryo, since the latter is not checked for).

Forest Reproductive Material seed, cones and parts of plants intended for the production of plants; also young plants raised from seed or from parts of plants.

Gamete sex cell (1N) capable of fusion with another gamete to produce a fertilized zygote (2N).

Gametophyte haploid plant (1N) which produces gametes (1N) by mitosis. the ability of an individual to pass on all its good (or bad) characteristic to all of its progeny.

Genetic qualities qualities derived from the inherited characteristics of an individual tree.

Genetic variability the presence in a species of different grades of the same characteristic which allows selection and breeding for that characteristic to be carried out.

Genetically improved possessing qualities derived from inherited characteristics, which qualities are superior to the average for a population as a whole.

Genotype an individual organism's hereditary constitution which may or may not be expressed phenotypically.

Germination growth of an embryo resulting in its emergence from the seed.

Gymnosperm woody plants; ovules not enclosed in ovary; xylem without vessels.

Gynoecium female part of the flower, made up of one or more ovaries, and their styles and stigmas.

123

Haploid
an organsim or cell having only one complete set of chromosomes, ordinarily half the normal diploid number (usually abbreviated as 1N). It is characteristic of gametes of vascular plants.

Hectare
metric unit for measuring land area 1 hectare = 2.47 acres.

Hectolitre
metric unit for measuring cones in some European countries; equal to 100 litres or 2.74 bushels.

Heritability
the ability of a particular characteristic to be inherited in subsequent progeny.

Hermaphrodite
containing both stamens and ovary.

Homogamous
the anthers and ovaries maturing simultaneously.

Hybrid vigour
the increased vigour often resulting from the crossing of two closely related but distinct species.

In-breeding
the repeated use of the same individual or close relative in a breeding programme.

Indigenous
of a stand of trees, native to the locality.

Inflorescence
flowering branch, or portion of stem above the last stem leaves, bearing branches and flowers.

Integument
the layer of tissue in conifers that encloses the nucleus of an ovule, and which develops into the seedcoat.
A protective structure that develops from the base of an ovule and encloses it almost entirely, except for an opening, the micropyle, at the tip of the nucellus.

Inter-fertile
the ability of individuals to breed with one another.

Inter-specific hybrid orchards
Seed orchards formed by growing two closely related but inter-fertile species together.

Internode
that part of a plant stem separating two distinct whorls of branches.

Intra-specific hybrid orchards
Seed orchards formed between two different origins of the same species.

Involucre
bracts forming a more or less calyx-like structure round or just below the base of an inflorescence or fruit.

Mast
colloquial term used to describe tree fruit crop; usually associated with broadleaves.

Megagameto-phyte
haploid (1N) nutrient storage tissues in coniferous seeds. This tissue is often mistakenly called the 'endosperm' in conifers.

Megasporangia
ovules.

Megaspore
first cell of the female gametophyte which eventually becomes the embryo sac.

Megasporophyll
modified leaf, or cone scale, bearing megasporangia.

Meiosis
a series of complex nuclear changes within the original cell, resulting in the production of new cells with half the number (1N) of chromosomes characteristic of the original cell.

Micropyle
small channel between the tips of the integument at the apex of the ovule, through which the pollen tube travels, prior to fertilization.

Microsporangia
pollen sacs.

Microspore
first cell of the male gametophyte which eventually becomes a pollen grain.

Microsporophyll
modified leaf, or scale, bearing microsporangia.

Mitosis
a process of precise duplication of genetic material in which the cell nucleus divides into two new nucleii, each of which has the same number (2N) of chromosomes as the original cell.

Monoecious
having both sexes on the same plant.

Native
Not known to have been introduced by human agency.

124

Normal seed stand	a seed stand in which between 25 and 50% of the dominant trees are healthy and have straight stems and good well shaped crowns. This term was used prior to the introduction of the Forest Reproductive Material Regulations 1973.	**Phenotypic qualities**	qualities resulting from the response of trees to the local site and environment.
Not-tested orchards (NT)	seed orchard in which the clones used have not been tested and shown to be superior for the chosen characteristic and to have good combining ability.	**Phytosanitary Certificate**	certificate issued as a result of a test carried out to ascertain that seed or plants are free from general or specific disease organisms not endemic in or desired to be introduced into another country. May only be issued after quarantine or disinfection. Also known as Health Certificate.
Nucellus	Nutritive tissue in an ovule.	**Pistil**	structure comprising stigma and style.
Nucleus	a dense, spherical, somewhat transparent, body in the protoplasm of plant and animal cells, which contains chromosomes.	**Placenta**	the part of an ovary to which the ovules are attached.
		Plus seed stand	see Almost plus and Normal seed stands. Over 75% dominant trees.
Nut	fruit with a bony, woody, leathery or papery pericarp; usually one-seeded and partially or wholly enclosed in a husk (involucre).	**Pollen/pollen sac**	microspores of angiosperms and gymnosperms; the structure on an anther containing pollen.
Origin	place in which an indigenous stand of trees is growing, or the place from which a non-indigenous stand was originally introduced.	**Pollen cone**	the male reproductive structure of conifers, which produces pollen grains. It consists of an axis bearing spirally arranged scales, each of which supports two pollen sacs containing pollen.
Ovule	a female organ surrounded by integument, within which an egg cell (1N) is produced, and which matures into a seed (2N) following fertilization.	**Pollination**	the transfer of pollen from the pollen cone to the receptive part of the seed cone.
Ovuliferous scale	scale-like part of the female cone on which the naked ovule is borne in conifers.	**Pome**	fruit consisting of an enlarged fleshy receptacle surrounding the pericarp.
Peduncle	the stalk of all or part of an inflorescence.	**Primordia**	rudimentary structures; structures at their earliest stages of development. The earliest identifiable tissues which subsequently develop into leaf, flower, etc.
Pericarp	outer structure of maturing fruit.		
Periodicity	the interval (in years) between good cone crops. Some trees in a stand, or area, may bear cones every year, but heavy crops are periodic, usually occuring several years apart.	**Progeny**	offspring of plants.
		Progeny tested	an individual shown by a series of tests to have the ability to pass on desired characteristics to its progeny.
Phenotype	all characteristics — morphological, anatomical, and physiological — of a plant, determined by the interaction between genotype and environment.	**Protandrous**	stamens maturing before the ovary.
		Protogynous	ovary maturing before stamens.

Provenance the place in which any stand of trees, whether indigenous of non-indigenous, is growing.

Purity the amount of pure seed of the specified species in a seed lot expressed as a percentage of the bulk seed.

Recalcitrant seed term used to describe a seed that does not obey the normal seed storage principles that the drier the seed and the cooler it is kept the longer it will maintain its viability. Seeds which conform to the principle are called orthodox seeds.

Region of Provenance for a species, sub-species or variety, the area or group of areas subject to practically uniform ecological conditions, in which are found stands showing similar phenotypic or genetic characteristics; for a seed orchard, the region of provenance of the material used for the establishment of that seed orchard.

Registration the act of placing a seed stand on the National Register of seed stands.

Reproductive Material see Forest Reproductive Material, Selected Reproductive Material and Tested Reproductive Material.

Samara dry indehiscent fruit, part of the wall of which forms a wing.

Seed a matured ovule containing an embryo and nutritive tissue enclosed by a protective seedcoat, which is capable of developing into a plant under suitable conditions.

Seed cone the female reproductive structure of conifers, which produces seeds. It consists of an axis that supports spirally-arranged bracts, with ovuliferous scales at the base of each bract. Two ovules, which become seeds after fertilization occurs, are attached to the upper surface of each ovuliferous scale.

Seed orchard specially selected collection of trees, planted in an orchard fashion, established to produce seeds, usually of improved genetic quality. Seed orchards may be clonal (i.e. propagated from scions and produced from grafts or rooted cuttings) or seedling (i.e. propagated from seeds). An orchard may be described as first generation (from untested [natural stand] parents) or as advanced generation (the offspring of superior parents selected from a genetic test plantation). Some orchards may be established to produce seeds of species that do not produce adequately in natural stands.

Seed planning zone in accordance with seed transfer rules, an area throughout which seeds of a given provenance may be transferred and in which the resulting seedlings can be expected to perform adequately.

Seed production stand a forest stand reserved and managed as a source of seeds.

Seed source the place (latitude, longitude, and elevation) at which seeds are collected. The source of a seed collection may not be identical with its provenance.

Seed zone a geographic area usually delimited by some natural boundary and showing nearly uniform ecological conditions in which trees of a given species are regarded as distinct from those in neighbouring zones. Widely used in North America in natural forest stands.

Seedcoat the protective outer layer of a seed derived from the integument of the ovule.

Seedling orchard an orchard laid down for seed production but produced from seedlings raised from seed and not clonally by vegetative means.

Seed lot a quantity of seeds of the same species, provenance, date of collection and handling history, and which is identified by a single number.

Selected Reproductive Material forest reproductive material (FRM) derived from basic material approved for registration under the FRM Regulations.

Selected seed stands	seed stands derived from seedling orchards.	**Stamen**	male reproductive organ of plant.
		Staminate	of such organs.
Serotinous	a term applied to cones that remain on the parent tree, without opening, for a year or more after the seeds inside have matured.	**Sterile**	of seed, not capable of germination; of stamens, not capable of producing viable pollen.
Silvicultural selection	selection by humans (as opposed to nature) of trees that should be removed to improve the form of a stand, such as thinning, pruning, etc.	**Stigma**	the receptive surface of the gynoecium to which pollen grains adhere.
		Strobilus	multiple-seeded fruit with catkin or cone-like structure, e.g. *Betula, Alnus*.
Source identified	category of seed certification available in the OECD scheme but not in the EEC which seeks to guarantee that the seed was actually collected in the area or seed-zone specified on the certificate.	**Supplier's Certificate**	Seed and Plant Suppliers Certificates are required by virtue of the Forest Reproductive Material Regulations to be issued within 14 days of invoice to the Purchaser. They must contain certain specified information and be issued on the appropriate coloured certificate.
Source identified sub class A	as above but daily observations are made to ensure seed is collected from a given zone.	**Testa**	skin or outer coat of a seed.
Source identified sub class B	as above but only the registration procedures are carefully checked; no daily field inspection takes place.	**Viability**	the presence in a seed of viable tissue which, provided the appropriate dormancy breaking .treatment is experienced, will lead to its germination when conditions permit.
Specific combining ability	the ability of an individual to pass on a specific good characteristic to all of its progeny.	**Zygote**	diploid cell (2N) resulting from the fusion of two haploid gametes (1N), the product of fertilization.
Sporangium	a structure containing spores.		
Sporophyll	a leaf-like structure bearing sporangia.		

Index

Trees, insects and fungi are indexed under their botanical or scientific names.

Turkey 45; *see also* origins, seed

Ulmus spp. 104
glabra 38
United States 4, 45-46, 102; *see also* origins, seed
Universal Decimal Classification 24-26, 28-46

Vancouver Island: *see also* origins, seed
variation 8, 9, 13-14
vegetative propagation 3, 5-7, 29, 50
vermin 119
viability tests 108, 110-113

Washington 45-46; *see also* origins, seed
weather 8, 71-72, 74-76, 78, 83, 116
White Russia 10
windthrow 4
wings, dewinging 91-92

X-ray techniques 79, 95, 110, 114

yields (of seed) 72-78, 90, 93-94, 116

Acknowledgement   This index was prepared by
Dr John Chandler.

Printed in the United Kingdom for HMSO
Dd 291255 C35 6/92